The BEST of
Sasquatch
Bigfoot

The Latest Scientific Developments

plus all of

On the Track of the SASQUATCH

and **Encounters with BIGFOOT**

John Green

hancock

house

ISBN 0-88839-546-9
Copyright © 2004 John Green

Second printing 2013

Cataloging in Publication Data

Green, John, 1927–
    The best of Sasquatch Bigfoot : the latest scientific developments
plus all of On the track of the Sasquatch and Encounters with Bigfoot / John Green.

Includes index.
ISBN 0-88839-546-9

    1. Sasquatch. I. Title. II. Title: On the track of the Sasquatch.
III. Title: Encounters with Bigfoot.

QL89.2.S2G728 2004   001.944   C2004-902309-8

Printed in South Korea — PACOM

Editor: John Green
Production: Rick Groenheyde, Ingrid Luters
Cover Design: Rick Groenheyde

We acknowledge the financial support of the Government of Canada through the
Canada Book Fund for our publishing activities.

Published simultaneously in Canada and the United States by

HANCOCK HOUSE PUBLISHERS LTD.
19313 Zero Avenue, Surrey, B.C. Canada V3S 9R9

HANCOCK HOUSE PUBLISHERS
1431 Harrison Avenue, Blaine, WA U.S.A. 98230-5005

(604) 538-1114   Fax (604) 538-2262
(800) 938-1114   Fax (800) 983-2262
Web Site: www.hancockhouse.com   *Email:* sales@hancockhouse.com

Contents

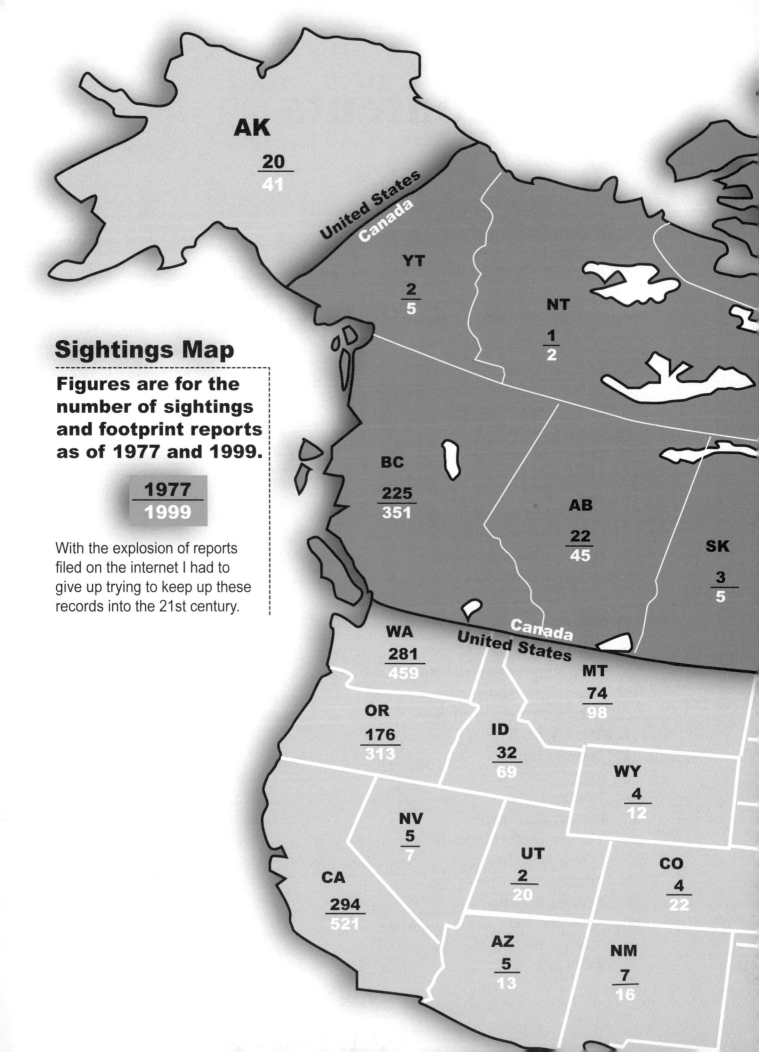

AK
20
41

United States
Canada

YT
2
5

NT
1
2

Sightings Map

Figures are for the number of sightings and footprint reports as of 1977 and 1999.

1977
1999

With the explosion of reports filed on the internet I had to give up trying to keep up these records into the 21st century.

BC
225
351

AB
22
45

SK
3
5

Canada
United States

WA
281
459

MT
74
98

OR
176
313

ID
32
69

WY
4
12

NV
5
7

UT
2
20

CO
4
22

CA
294
521

AZ
5
13

NM
7
16

The Ape and the IM Index

The first four chapters of this book deal with significant events in the Sasquatch/Bigfoot investigation in the twenty-first century, but the others are a reprinting, without updating, of two books, *On the Track of the Sasquatch* and *Encounters with Bigfoot,* that in various versions have been continuously in print since 1968. Paradoxically, it is the fact that I wrote much of them so long ago that makes them uniquely relevant in 2004.

Although the story is actually much older, reports of outsize humanlike footprints and huge upright-walking animals first attracted attention in 1958 when a cast of a "Bigfoot" print was made and publicized, and became more widely known in 1967 when a man named Roger Patterson took a 16 mm movie purporting to show one of the creatures.

In those days such reports made news, but in recent years new evidence for the existence of the sasquatch is usually ignored by the media. Proof that a bipedal ape shares this continent with humanity is apparently considered so big a story that it can't possibly be true. The headlines now are reserved for stories of the opposite kind, claims of proof that the Bigfoot tracks

and the movie were just fakes after all. Not many people were involved in investigations on site either in 1958 or in 1967, and only two people took part in both, the late Bob Titmus and myself. Bob never wrote of his experiences, so my books are the only equivalent of that courtroom staple, the investigator's notes made at the time.

At the end of 2002 newspapers and TV networks all over the world had a field day with a yarn that all the footprints were faked by a man who had just died, so Bigfoot was also dead. Even though the story was obvious nonsense its effects will last a long time, stopping witnesses from risking ridicule by making their stories public, and discouraging scientists who might be considering getting involved in the investigation. That fiasco is dealt with in full in a later chapter.

As for the movie, attempts to debunk it come along every year or so, usually contradicting each other. It became widely believed in Hollywood that the man who changed the faces of the actors in the *Planet of the Apes* movies also created the creature in the Bigfoot film. He apparently never denied it while he was working, but after he retired he told sasquatch investigator Bobbie Short, on tape, "I was the best but I wasn't that good!"

The Patterson film.

The BBC film recreation.

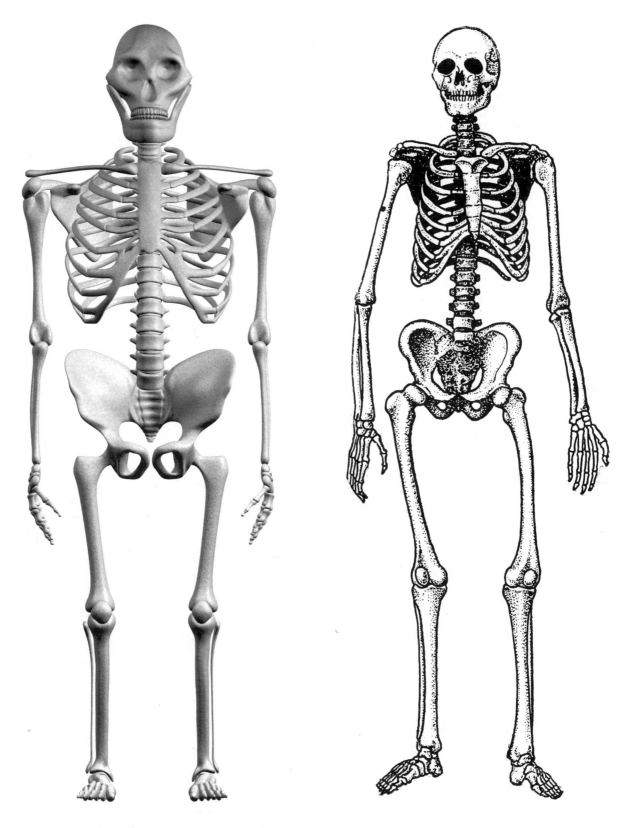

On the left is the illustration of the skeleton of the creature in the Patterson movie and on the right is an illustration of a normal human skeleton.

The illustration of the sasquatch skeleton was created by Reuben J. Steindorf, a forensic animator at Vision Realm Entertainment Inc. They first converted 116 frames of Roger Patterson's Bigfoot movie into pixels at 72 dots per inch. Then using sophisticated animation software their computer was able to determine the pivot points of the joints as it tracked the movements of the creature from frame to frame. This process established the relative length of the limbs with a high degree of accuracy, and confirmed previous observations that the width of the creature compared to its height, and the length of its arms compared to the length of its legs, take it well outside the range of human measurements.

The prestigious Wildlife Unit of the BBC also took a hand in the debunking game. They succeeded in making themselves look foolish by showing a pitiful attempt at a re-creation of the Patterson movie with a man in an ape suit, and by claiming as proof of fakery a copy of a letter dated after the movie was made which indicated that Roger made money by selling rights to show it. Who wouldn't? Another wing of the BBC had been one of the earliest to ante up.

Why anyone would argue that selling something of value proved that it wasn't genuine is hard to understand.

More recently, a book was published in which the author claimed to have found the man who wore the ape suit in the movie, and the man who sold the suit to Roger Patterson. In each case there was no evidence, just one person's story, and the two men described two totally different suits. The man who claimed to have worn the suit said Roger had made it by skinning a dead horse. It was in three pieces and it stunk. The man who claimed to have made the suit said it consisted of six pieces and was made of modern materials.

Paradoxically, this silly attempt to prove that Patterson hoaxed his film led to the discovery that the movie itself has always contained proof that it does not show a man in a suit.

One of the things that the supposed suit maker is quoted as saying is that the way to make the arms in the suit look longer than human arms is to extend the gloves of the suit on sticks. Many people have noted that the arms of the creature in the film look unusually long, almost as long as its legs. Some, including myself in 1968, have published estimates of their length (see page 77). No one went on to deal with the question of how human arms could be extended to match the extra length and what such an extension would look like.

There is no way to establish for certain if any of the dimensions estimated for the creature in the film are accurate, but what can be established with reasonable accuracy is the length of the creature's legs and arms in relation to one another. From that ratio it is simple to calculate how many inches must be added to the arms of a man of known size in order to make them long enough

The arm and leg bones of the creature in the Patterson film as determined by Reuben J. Steindorf of Vision Realm Entertainment Inc. from their computer study of 116 frames of the film.

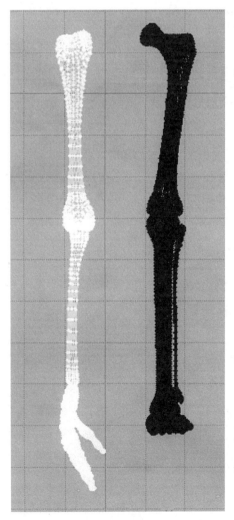
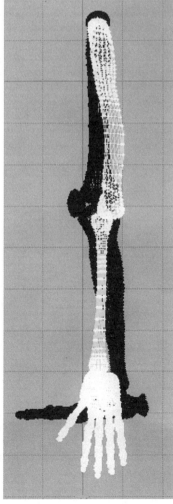

to fit the supposed suit. In my own case the answer turned out to be about 10 inches.

But in order for the arms to bend at the elbow, which they plainly do in the movie, all of that extra length has to be added to the lower arm. The result, in my case, is about 12 inches of arm above the elbow and 29 inches below it—an obvious monstrosity. The creature in the movie has normal-looking arms. It cannot be a man in a suit!

Many issues in the long debate about the movie remain unresolved—what the film speed was, whether a man could duplicate the creature's unusual bent-kneed walk, whether its behavior was normal for an animal, whether the tracks left on the sandbar could have been faked, and so on—but all of them turn out to have been irrelevant to the main issue.

My measurements of the film, made 36 years ago, gave the creature arms that were 30 inches from the shoulder to the wrist and legs that were 35 inches from the hip to the ground. My own measurements are about 24 inches from shoulder to wrist and 40 inches from hip to ground. Scientists studying primates use almost identical measurements, the only difference being that they measure to the ankle joint rather than the bottom of the foot, to establish what is called the intermembral index, which is one of the things used to distinguish one kind of primate from another.

Gorillas and chimpanzees, with arms longer than their legs, have average indices of 117 and 107 respectively. The average human IM index is around 70. Only the ratios of the measurements matter, actual size makes no difference. Establishing an accurate IM index for the creature in the film is difficult, since no one frame shows all of both the upper and lower limbs at right angles to the camera, but it can be done, in fact a computerized study of the creature's walk done for the TV documentary *Sasquatch, Legend Meets Science* has already done it. Using sophisticated forensic animation software to follow points on the creature's body and limbs as it moves through 116 frames of the movie, the computer was able to produce pictures of its skeleton showing an IM index between 85 and 90.

Forensic animator Reuben Steindorf's comment after studying the film was that making it using a man in a suit would require a lot of mechanisms not available in the 1960s. It would have to have been a highly funded project and there would have to have been trailing electric cables attached to the creature somewhere. In short, it couldn't be a man in a suit.

A study of a lesser number of frames by Dr. Jeff Meldrum, an anatomy professor at Idaho State University, produced a similar result, and he also noted that besides bending its elbows the figure in the film flexes both its wrists and its fingers, "all but ruling out the possibility that an artificial arm extension could be involved."

It will no doubt take a while before the impact of the IM index makes itself felt among primatologists, but they can hardly ignore one of their own standard measurements when it tells them that there really is a giant higher primate to be found in North America.

Recent Developments

Since *On the Track of the Sasquatch* was last revised there have been huge changes in the overall picture. In the late 60s I was in touch with almost all of the few people who were investigating this subject, and all of us together probably knew of less than 100 sighting reports. After Roger Patterson's film caught public attention a lot more reports began to come to light, until I was recording about 100 sightings or footprint finds each year. Still we always suspected that the great majority of incidents never became generally known because most of the people involved did not know of anywhere to report them without being ridiculed.

With the growth of the internet that situation has turned upside down. There are many websites where people are asked to submit such reports, anonymously if they choose. I don't know of anyone who tries to monitor all the sites, but Matt Moneymaker's Bigfoot Field Researchers Organization alone gets at least half a dozen reports a day. A lot of them are obviously from pranksters, and trying to sort out the less-obvious fakes from the genuine information is a major task, but what remains must be more than a thousand reports a year. For a dozen years I worked at getting all the information I had into a computer database and by the spring of 2001 I had worked all the way through my back files and had more than 4,000 entries, but by then it was also obvious that I could no longer keep up with the data that was available on line. The new reports have cleared up one anomaly. From the state of Colorado, with a sea of mountains and a hockey team that displays Bigfoot tracks on the shoulders of its uniforms, I did not record a really substantial report in 40 years. More recently many reports have surfaced there, several of them among the best from anywhere. There is a somewhat similar situation with reference to the province of Alberta, except that most of the new information comes not from the internet but from years of dedicated investigation by Tom Steenburg, author of *The Sasquatch in Alberta* and *Sasquatch: Bigfoot The Continuing Mystery.*

The most interesting thing about the flood of new information, however, is that the majority of the reports do not come from the traditional areas at all. There are far more reports from east of the Mississippi than there are from the west of the continental divide. I have done enough investigating to satisfy myself that the evidence from the Midwest, East and South is on a par with what I am familiar with in the West, but reports from those areas are not the subject of this book.

The other huge change is in the attitude of some of the scientists. For many years Dr. Grover Krantz was the only physical anthropologist willing to gamble his career by publicly being a full participant in the sasquatch investigation, and there were no zoologists involved at all. The small group that gathered for the first viewing after Roger Patterson got his remarkable movie in the fall of 1967 did not include anyone with scientific credentials.

It was a different story in the fall of 2000 when the BFRO organized a group effort at a place called Skookum Meadows east of Mount St. Helens and brought back evidence perhaps equal in importance to the Patterson movie: a huge cast made where a large animal had left limb and heel prints in a mud patch. One of the three men who found the print was a zoologist, Dr. Leroy Fish, and among the five additional people assembled when the cast was being cleared of its coating of dirt I was the only one without a doctor's degree. The impression was found where the men had placed some fruit at night in the middle of a patch of soft mud surrounded by mud that had already dried hard. They were hoping to get footprints if a sasquatch was attracted to the fruit. When they returned a few hours later the fruit had indeed been disturbed, but instead of footprints what they found was a set of large, shallow depressions showing hair patterns, and a variety of holes in the mud. Some of the holes were identified as elk and coyote tracks, others were a puzzle. It took a while for the men to come to the conclusion that a sasquatch had sat down at the edge of the soft mud, leaving the impressions of slightly more than half of its buttocks and one thigh plus several prints where it had moved its heel around, and had leaned over onto a forearm as it reached across with the other arm towards the fruit.

Successfully casting all of such a large impression would normally have been out of the question, but one of the three men, Rick Noll, was a professional cast maker as well as a long time sasquatch investigator, and had with him a couple of hundred pounds of exceptionally strong plaster. Using aluminum tent poles for bracing, they made what became known as the "Skookum cast," preserving all the evidence except some apparent scratch marks near the fruit.

As with the original impression in the mud, the significance of the cast is not obvious at first glance, except for the humanlike heel shapes sticking up from it. Plainly something large and hair-covered had set itself down in the mud, but there are elk in the area and elk tracks in the cast. Careful examination, however, rules out all the common animals. Certainly no part of an elk could match the obvious Achilles tendon of the best heel print.

Professor Jeff Meldrum from Idaho State University, a physical anthropologist whose special study is the evolution of bipedal walking, took on the job of cleaning up the cast. He spent several days meticulously picking away the dirt adhering to it and in the process collected a lot of pieces of animal hair, but only a very few of them proved to be interesting. The most important thing he was able to do was to determine the location of the joints in the thigh and forearm impressions, which showed the bones to be half again as long as those of a six-foot man.

Besides Dr. Meldrum, I am in close touch with two other scientists who are publicly committed to the sasquatch investigation, zoologists Dr. John Bindernagel and Dr.

Henner Fahrenbach. Dr. Bindernagel, with 30 years field experience in many parts of the world, set out to determine if what sasquatch witnesses reported added up to a believable animal. He found that it did. Further, as noted in his book *North America's Great Ape, the Sasquatch*, he learned that some seemingly unlikely behaviors the witnesses described are shared with one or other of the known great apes. Significantly, some of those shared behaviors turned up in sasquatch reports before they were observed by scientists studying the other apes.

Dr. Fahrenbach did a statistical analysis of the footprint dimensions in my computer database and found that when plotted on a graph they form the normal bell curve that would be expected of a species of real animals. He has also specialized in the study of hair, and has found a number of suspected sasquatch hairs from widely divided locations that don't match hairs from known animals but do match each other. Unfortunately the hairs have so far failed to provide suitable material for DNA identification. Jeff Meldrum had earlier laid to rest a concern felt by some laymen like myself that someday an expert in foot anatomy would demonstrate that supposed sasquatch tracks showing long toes and those showing short toes could not both be genuine. Instead, after examining all the track casts and photos he could locate, he determined that not only were the tracks consistent anatomically, they also showed an ability to bend in the middle that human feet (and one-piece wooden feet) do not have. Better yet, he found that the tracks dictated a style of walking different from that of humans, but exactly like that of the creature in the Patterson movie. He has also continued the investigation initiated by Grover Krantz of fingerprint-like skin ridges found on a few footprint casts made in particularly fine material at widely-separated locations.

Skin ridges of this sort do not occur on the feet of animals other than primates. They might be considered a non-slip surface for tree-climbers that have no claws. This work recently caught the attention of a police fingerprint expert from Texas, Jimmy Chilcutt, who had made a study of such "dermatoglyphics" on the feet of humans and apes. Thinking he could prove that the footprint casts were faked, he examined them, found that they showed the same unique pattern, and pronounced that they were proof of the existence in North America of an unknown great ape.

Also on the scientific front, and ironically at the same time that the media were going ape over the yarn that Bigfoot had died with Ray Wallace, a really important story was published in one newspaper, broadcast on one cable network and totally ignored by everyone else. That story was about prominent figures in zoology and anthropology who are now saying that the sasquatch evidence deserves serious study.

On January 5, 2003, the *Denver Post* devoted several pages, including half the front page, to stories by Theo Stein, their environment writer, in which he quoted a series of primate experts giving their stamp of approval:

As far as I am concerned the existence of hominids of this sort is a very real possibility:
— Dr. Jane Goodall, world famous for her studies of chimpanzees.

There have been so many sightings over the years. Even if you throw out 95 percent of them, there ought to be some explanation for the rest. The same goes for some of these tracks:
— Dr. George Schaller, director of science for the Wildlife Conservation Society and first scientist to do a major study of wild mountain gorillas.

I think a serious scientific enquiry is definitely warranted:
— Dr. Esteban Sarmiento, primate specialist at the American Museum of Natural History.

I'm not one to pooh-pooh the potential that these large apes may exist:
— Dr. Russell Mittermeier, president of Conservation International and chairman of the Primate Specialist Group.

It's not conclusive, but it's consistent with what you'd expect to see if a giant biped sat down in the mud:
— Dr. Daris Swindler, author of *An Atlas of Primate Gross Anatomy*, commenting on the back of a heel and part of the Achilles tendon shown in the Skookum cast.

Unlike the Ray Wallace story, this one was not picked up by the Associated Press and no word of it reached readers of other newspapers.

Also in January, 2003, the U.S. Discovery Channel aired a one-hour documentary produced by Doug Hajicek of White Wolf Entertainment and titled *Sasquatch, Legend Meets Science.* The show included footage of all the people mentioned above except Drs. Goodall and Mittermeier. It also called on the expertise of specialists in animation to study the gait of the creature in the Patterson movie, which they determined did indeed have a very non-human way of walking. It was also noted to have a bump that rose and fell on its right thigh which appeared to indicate a type of hernia sometimes suffered by human sprinters.

In the documentary Dr. Swindler went well beyond the statement quoted in the *Denver Post*, saying:

In my opinion the impression is not made by a deer, a bear or an elk nor was it made artificially. The Skookum body cast is that of an unknown hominoid primate.

Like the *Post* story the documentary was not mentioned by other media, but unlike the story it will continue to be shown on television and is available in other forms.

Pete Travers sketch of possible sasquatch position as it took fruit placed in the muddy road by researchers at nearby Skookum Meadows.

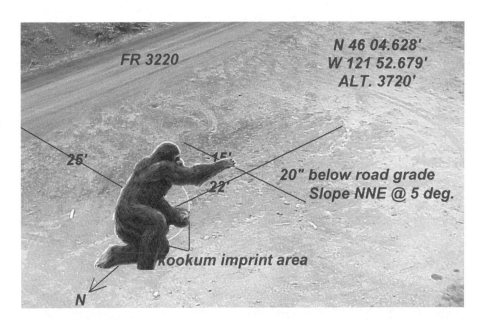

Rick Noll's cast of the heel print and Achilles tendon of a 200-pound man (himself) beside the best heel print in the Skookum cast.

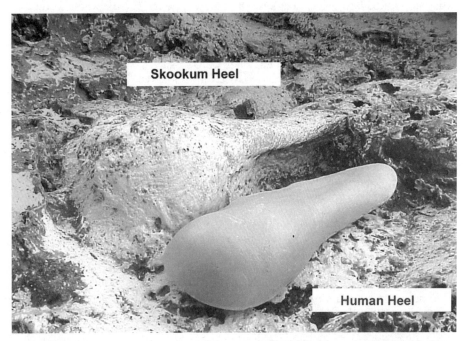

Dr. Jeff Meldrum, the physical anthropologist who conducted the most detailed study of the Skookum cast pointing out the position of the various body parts.

Big Foot Did Not Die

After the death of a man named Ray Wallace in December, 2002, surviving family members told the reporter who wrote his obituary that "Bigfoot" had also died. Ray Wallace, they said, was Bigfoot. He, not a huge, unknown animal, had made the big tracks that were first reported in 1958 on and around a dirt road his company was pushing through the Bluff Creek valley in the northwest corner of California. Family members agreed that they had always known about it and that Ray did it as a joke on his employees, walking around wearing a huge pair of carved wooden feet. For proof one of them showed a photographer just such a pair of carved feet, with strap harness attached.

The story was nonsense on the face of it, since everyone who had looked into the subject knew that huge bipedal tracks had been reported from all over the North America starting long before Ray Wallace was born. No matter, apparently it was just the sort of tall tale many editors were waiting to see and eager to tell. They were so sure the whole Bigfoot phenomenon had to result from fakery that they rushed into print and on the air proclaiming to the world over and over again that the whole Bigfoot thing was just one man's hoax.

Of more than 50 papers that spread the story, and even more radio and TV stations, not one bothered to check its accuracy. Apparently none of them realized what the tracks in question were actually like, and they had no interest in finding out. Had even one of them bothered to learn all that was involved and then asked the Wallaces to show that they could duplicate it all by walking around wearing the wooden feet, it would have killed the story they were having so much fun with.

I am not saying it can be proved that those tracks were not faked, or that Ray Wallace could not have been involved, but proving that many of the tracks could not possibly have been made in the way the Wallaces described would have been easy.

As the tale spread it got even more nonsensical. The Wallaces had said, as just about every sasquatch investigator already knew, that Ray had made fake Bigfoot photos and movies, featuring his wife wearing a fur costume. But they also said that Ray had nothing to do with the famous Bigfoot movie taken by Roger Patterson at Bluff Creek in 1967. The media firestorm, however, eventually made Mrs. Wallace the subject of the Patterson movie, with Ray as the cameraman. One story also had Ray sending younger members of the family as far away as British Columbia equipped with other carved wooden feet, to make all the big footprints ever seen everywhere. And to give an aura of authenticity many of the stories called it a "deathbed confession," although the obituary made no suggestion of any such thing. It was clearly presented as something his survivors had long believed but not as a dying claim by Ray himself.

That was strange enough, but what was stranger still, the media became so caught up in shouting that the Bigfoot hoax had been exposed that they would not allow any other voice to be heard. In two months not one newspaper would so much as have a reporter talk to someone, namely me, who told them that he had investigated the original incidents back in 1958, and had ample proof that Ray Wallace and the wooden feet could not have been responsible.

To give readers a sample of what kind of story teller Ray Wallace was, here, in part, is a message that I e-mailed to many of the newspapers that had printed the story:

"So Ray Wallace supposedly told his family that he created 'Bigfoot' by walking around in California with a pair of huge carved wooden feet, and his family supposedly believed him. And the media has now told the whole world that 'Bigfoot' was just Ray Wallace. Alright, but shouldn't the world also know what Ray Wallace has said of some of his other achievements ?"

Big Foot used to be very tame, as I have seen him almost every morning on the way to work. I would sit in my pickup and toss apples out of the window to him. He never did catch an apple but he sure tried. Then as he ate the apples I would have my movie camera clipping off more footage of him. I have talked to several movie companies about selling my movies which would last for three hours. The best offer I've had so far is $250,000.
— Ray Wallace letter to the *Klam-ity Kourier*, Oct. 1, 1969.

The first day we went out in search of the Mt. St. Helens apes we saw five different-sized tracks. The first day out from our camp we saw where five of the giant-sized apes had crossed a small creek, the water was still muddy in their tracks. We found the ape cave. I sent my pack crew out after a hundred pounds of plaster of Paris and I made some of the nicest casts of those Mt. St. Helens apes tracks that I have ever saw. I don't think I could ever find the ape cave that my guide showed me where the Mt. St. Helens Apes have stayed for possibly several thousand years.
— Ray Wallace letter to John Green, Feb. 6, 1967.

Please send me your correct address. I want to send you a picture of one of the male Mt. St. Helens apes that the loggers took this spring as they were feeding apples to an old pair of BFs and the female was carrying a baby, but she never came close enough for them to get a good picture, they got some close up pictures of this 9 foot tall male, I just borrowed the negatives. I want to send all of the BF researchers a picture.
— Ray Wallace letter to John/Green, Dec. 2, 1984.
[I sent the address but haven't seen any pictures — nor has anyone else!]

I sent you a tape of the songs about Big Foot plus some of his high-pitched screams, I would like for you to set up a meeting with scientists from all over the world to listen to those screams. Our government thinks these Big Foots are being let out of flying saucers.
— Ray Wallace letter to John Green, Oct. 20, 1989.

I just want to inform you Big Foot hunters that Big Footed creatures are people, they speak a language. I could tell you more about the Sasquatch or Bigfoot than anyone else, I told Roger Patterson where to go in California to see Big Foots. I made ten thousand feet of movies of the Big Foots before I told Roger Patterson where to go. I could take you to a cave in Northern California where the Big Foots live in a very rich gold mine cave. Did you know that Tom Slick bought Big Foot skeletons for many years and turned them over to the Pentagon in Washington, D.C.? legs bones four inches diameter, two and a half feet long between the ankle and knee I have talked to the Big Foots many times they didn't understand me and I didn't understand them, but their brown eyes told the story that they are very sad because the bear hunters are killing all their people.
— Ray Wallace letter to Dennis Gates, May 24, 1978.

Some of you Big Foot readers probably wonder how I got the Big Foot scream on tape.. in 1958 before the bear hunters got to chasing Big Foot with their hounds and made him so wild, I used to see one of the Big Foots almost every morning, eating elder berries along the road. I have seen Big Foot several hundred times. I didn't ever tell anyone about seeing those large, hairy type creatures as I was having a hard enough time keeping men on the job. Most of the men would quit the first time they saw those huge shaped human bare foot tracks. Then I would; have to start looking for a new crew I lost $40,000 on that road job. After having two of the Big Foots captured and getting loose, I have always said not to underestimate the great strength of old Big Foot.
— Ray Wallace letter to the *Klam-ity Kourier*, March 25, 1970.

Back in 1947 when I had my logging crew on a free moose hunting trip to Canada near Vanderhoof, B.C. we saw a family of six Sasquatches and they were as interested in us as we were in them. I have seen the Big Foots in Northern California and around Mt. Hood in Oregon and around Mt. St. Helens and they all look alike so I know that Big Foot and Sasquatches are all brothers or sisters. The largest BFI have ever seen at least four feet across the chest and very large arms and carrying a large round rock in each hand. I have a movie of one throwing a rock and killing a deer.
— Ray Wallace letter to University of British Columbia, Jan. 26, 1981.

Everyone says who has heard Big Foots screams in northern California, before all the Big Foots were killed and hauled down the Klamath River in a tug boat and out into the ocean 12 miles to where was a small ship anchored in international waters and frozen into a block of ice and then transported to Hong Kong and sold, so now there aren't any more left in northern California. or is there if they are being let out of flying saucers.
— Ray Wallace letter to John Green , April 15, 1979.

No editor anywhere printed that information, so I tried a different approach pointing out by e-mail and in some cases also by phone that the claims made in the obituary were a far more successful hoax than any that Ray Wallace had ever carried out while he was alive, and outlining the items of physical evidence available to prove that the story was nonsense. No newspaper would discuss that either.

Demonstrations were put on for both CNN and FOX News in which their own people, walking on fiberglass copies of genuine 15" tracks, learned that deep tracks can't be made in firm sand that way, even, in the case of CNN, by two men weighing a combined 440 pounds. The TV people who actually tried it and photographed it were quickly convinced, but in each case when the brief news segments were broadcast that was not mentioned and the sasquatch researchers were ridiculed for refusing to recognize the reality of the Wallace claims.

Finally, in an attempt to reach the public with an account of the true situation, the museum at Willow Creek, California, which has on display casts of many of the tracks involved, offered $100,000 for the first person who could demonstrate how humans could have faked them. That story was sent to 800 editors but except for a few local papers that were approached directly the media ignored it completely.

There was actually nothing new about such an offer except the amount. A $1000 challenge had been issued on TV back in 1958, with no successful takers. I have had a $5,000 offer in print for the past 25 years, with no one even enquiring. The fact is that in the 45 years since the original "Bigfoot" story broke, no one has ever been able to demonstrate how the tracks could have been faked.

Where does Ray Wallace fit in? The men who saw the tracks were employed by his company, but he was seldom there. He was based at Willow Creek, a couple of hours from the Bluff Creek project, but his friend Ed Schillinger recalls that he was usually away somewhere trying to drum up future contracts.

Ray did have quite a reputation as a practical joker. Speculation that he had a hand (or foot) in making the tracks surfaced early on, and was by no means ignored, but on investigation was dismissed as being impossible and silly. The problem was to figure out how anyone could have made the tracks, something that hasn't been done to this day. Ray himself issued outraged denials, insisting, as was only common sense, that monstrous footprints showing up on his worksite were disrupting the job and costing him money. Some time later he apparently developed a yen to share the attention Bigfoot had stirred up and began to spin his outrageous yarns. Later still, probably after he had moved back to his old home in Washington State, he began making and selling obviously fake casts. I used to see them at a lodge on Mount St. Helens which also sold my books. One thing that he never did, at least in public, was to claim that he had made the Bluff Creek tracks. Had he done so he would, of course, have been called on to prove he could do it.

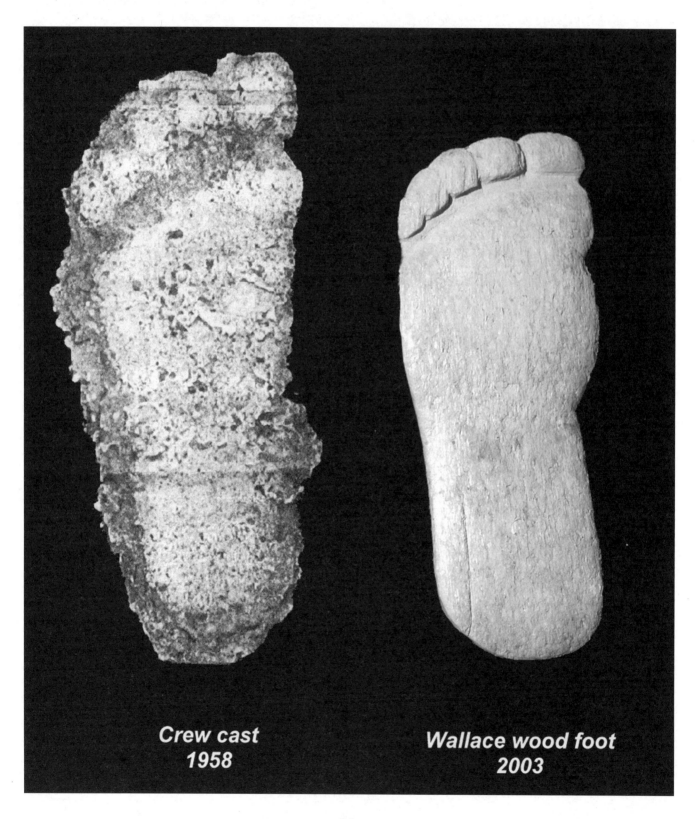

**Crew cast
1958**

**Wallace wood foot
2003**

What about the wooden feet that the current generation of Wallaces have displayed? So far there is nothing to show when in the past 45 years they were made or by whom, and none of them match the shape of the original "Bigfoot." The best pair does match the 15-inch track found later in 1958 on a sandbar in the creek and cast by Bob Titmus. They are somewhat crudely carved, and presumably they were made in imitation of those

**Titmus cast
1958**

casts. For them to be accepted as the originals with which the tracks were made someone would have to demonstrate how they could make imprints an inch deep in hard-packed sand and make deep, rounded toe impressions with their shallow, square-carved toes.

Were those or any of the other fake feet the Wallaces have shown ever used to make tracks that anyone accepted as genuine? It is certainly possible. This could have been done in soft mud, dirt or sand. Trying to match deep tracks in firm materials by wearing big wooden feet, however, is like trying to do it wearing snowshoes. People who do know some of the problems involved and yet would like to believe that the tracks were faked have come up with some really far-out suggestions: the depth was achieved with false feet mounted on tractor tracks; heavy concrete feet were hauled up and down with logging cables to make tracks on the steep slopes; the long strides were made by hanging onto the back of a moving truck; Ray Wallace faked the tracks of a monster because he wanted to get out of his contract so he was trying to scare his men into abandoning the job.

The media obviously believe that possession of big fake feet that can be worn is proof that the owner has used them to perpetrate a hoax, but most of the people I know who have made them, including myself, had the opposite idea. They were made to find out what could be done with them and what could not, and what fake footprints made with rigid, carved feet would look like. And did Ray really tell his family that he had faked all the tracks? I had a reason not to question their claim that he did. I was once present when another notorious yarn spinner told his children and grandchildren an equally outrageous tale. I am told, however, that later his son admitted that Ray had never actually said it, they just assumed it. As to his claim that he told Roger Patterson where to go to get his movie, a description he included in a letter to another researcher made it clear that Ray did not even know where that place was or what it looked like.

Aside from these introductory chapters, the new printing of *On the Track of the Sasquatch* is not a revision. It tells the story of things that came to light in investigations from the mid 1950s to the late 1970s, including a good deal that had happened long before that time. It does not begin with "Bigfoot," and it makes no mention of Ray Wallace, for the simple reason that until the media decided in 2002 that he alone was "Bigfoot" there was no reason to mention him. The following chapters will make clear that long before 1958 giant hairy bipeds and their big footprints were already an old, old story, particularly about the "sasquatch" in Canada and the "apes" of Mt. St. Helens in Washington, and that footprints almost identical to those cast at Bluff Creek in 1958 had been studied and cast in British Columbia in 1941.

$100,000 Reward

Here is the text of the Willow Creek museum offer:

$100,000.00 for BIGFOOT TRACKS

One hundred thousand dollars is being offered by the Willow Creek China Flat Museum for anyone who can demonstrate how the "Bigfoot" tracks that were observed in the Bluff Creek valley in northern California in 1958 and later could have been made by a human or humans.

This offer is genuine. It is not a joke or a publicity stunt. The money has been arranged for, and the first person or group who can meet the conditions of the offer will receive it. Everyone should understand, however, that the conditions are not easy. The offer is a direct result of recent publicity which has created a perception that the Bluff Creek tracks were just a hoax carried out by a practical joker walking around wearing a large pair of carved wooden feet, but it is not meant as a challenge to the people who originated that story, who may well be perfectly sincere.

The offer also is not a prize for technological achievement, such as being the first to build an effective footprint-stamping machine. It relates entirely to the question of whether the real tracks which brought the "Bigfoot" phenomenon to public attention could have been made by humans under the real conditions of the times and the places in which they appeared. The museum has casts of some of the tracks concerned, a few of them copies but mainly originals, available for inspection. It also has some related photographs, and published accounts of what was done and observed in connection with the tracks. There are also people still available for consultation who studied the tracks when they were made.

A formal document setting out the requirements to qualify for the award will take time to prepare, but a successful applicant will have to be able to make flat-footed, humanlike tracks with more than twice the area of human feet and longer-than-human strides which do the following:

> *Traverse a variety of terrains, including climbing, descending and crossing steep slopes covered with underbrush;*
>
> *Show variations of shape and toe position and stride accommodating to the terrain;*
>
> *Sink into firm ground to far greater depth than human footprints specifically as much as an inch deep in hard sand where human prints barely penetrate at all;*
>
> *Leave hard objects in the ground, such as stones, sticking up above the rest of the track.*

The applicant will also have to be able to make these tracks under the following conditions, although not all in combination:

> *In the dark, hundreds in a single night;*

> *In places where it is impossible to bring any vehicle or other machine or any equipment except what humans or animals could carry;*
>
> *Without doing anything to attract the notice of people a few hundred yards away.*

The reason that full specifications could not be included in the announcement was that the museum hoped publicity about the offer would lead to contact with the former roadbuilders and others who had seen tracks in the 50s. They were needed because it appeared that aside from myself everyone who had gone there to investigate the tracks in 1958 and 1959 had since died. As the chapters in this book titled "Bigfoot at Bluff Creek" and "Blue Creek Mountain" will explain, I had seen a great deal of a 15-inch, differently-shaped track that was found in the fall of 1958, but little of the original 16-inch tracks on the road project.

Some record of what the tracks were all about is available in the newspaper files of the day:

WILLOW CREEK - Bigfoot has been a familiar character to this part of the world off and on over a period of years.
> — Eureka, California, *Humboldt Times*, Oct. 7, 1958.

In soft places the prints were deep, suggesting a great weight.
> — Jerry Crew, *Humboldt Times*, Oct. 5, 1958

He described the tracks as being pretty heavy.
Quoting Julian Paulus regarding big tracks seen on a road job near Korbel in the spring of 1958.
> — *Humbolt Times*, Oct 5, 1958.

The footprint looks human but it is 16 inches long, seven inches wide, and the great weight of the creature that made it sank the print two inches into the dirt. Crew says an ordinary foot will penetrate the dirt only half an inch.
> — Associated Press story from Eureka, Oct. 6, 1958.

Judging from the deep indentation of the tracks he must be somewhere between 400 and 500 pounds. He must be quite an agile fellow leaping logs at a single bound and tracking throughout the wilderness covering a large territory quickly.
> — Edward Van Schillinger, stake setter on the Bluff Creek Road project, *Humboldt Times*, Oct. 8, 1958.

The first actual line of tracks definitely jolted me. On the hard ground where Philip Ammons' number 12's made a very light imprint, the track of Bigfoot sunk a half, to three quarters of an inch in depth. Twenty clear deep footprints marched along the side of the traveled portion of the road. Eighteen more were

seen at intervals where the trucks had not run over them. We followed them down the road for some distance and found them in both hard and soft earth.Bigfoot's tracks are in perfect proportion to what one would expect in their stride of sometimes 60 inches, 52 inches or the one short step over a small mound of dirt which was 40 inches.
Even the depth to which the track had been pressed into the ground was in keeping with their size.
— Willow Creek correspondent Betty Allen, in the *Humboldt Times*, Oct. 9, 1958.

We thought he might weigh as much as 400 pounds. He made firm footprint in the hard ground. Measuring the footprints for a distance of more than 60 feet we found the average stride to be 50 inches. We checked this against the stride of a man 6 feet 4 inches tall, with long legs and his stride was 30 inches. We were told by people who saw footprints made when this unknown man was running that they were 10 feet apart. He does his traveling at night. We learned these tracks have been appearing for the past 10 years.
— Seattle taxidermist Al Corbett quoted in the *Humboldt Times*, Oct. 1958.

They say, (the source of authority who isn't sure but talking) that the tracks are made by spacing carved feet a certain distance apart on the treads of a tractor, or on a roller used to smooth the road. Is that possible?
Individual measurements show some tracks to be sixty inches apart, some fifty-two inches and others 40 inches apart. Here and there, they show on one side and the other, some times as a small mound of dirt. Sometimes the tracks step easily up or down rough terrain. It is not necessarily in the path of a roller. In other places the tracks are within inches of the edge of the road in others in varying distances from the oiler rig or trucks. The ground may be that which tractors have run over. Sometimes the surface is perfectly smooth. The weight of the entire foot varies in depth, and according to the surface on

which Bigfoot has been walking. It doesn't respond to the "mechanical" explanation. The case of the wooden feet that "they say" are in existence if true they must be magnificent models of workmanship. Each toe is separate, tiny lines of the human foot are visible.
Then one asks if the toes are hinged to give the startling realism of action observed in the big tracks there are those who answer with a "yes." On Thursday morning the latest evidence debunks a lot of "mechanical" claims. That morning, the big tracks of Bigfoot were observed plunging down the side of a hill in the roughest of shale. The huge dug in (sic), the weight caused the feet to slide. What a way to treat someone"s carefully developed mechanical handiwork? There is $1,000 (over $10,000 in today's money) which could go to the fund for the badly needed hospital project of the Community Health Association at Hoopa if the "wooden feet" could be located, proven to be wearable, to produce Bigfoot's tracks. So far, the quest for them has been as fruitless as Coronado's search for the famed "Seven Cities of Cibola".
— Betty Allen, Willow Creek correspondent, in the *Humboldt Times*, Oct. 31, 1958.

Crossed the creek and there on the other side were the huge prints going upstream however he seemed to have been just snooping around when the tracks were made up and down banks, in and out of the timber and underbrush, down the creek and back, over huge boulders,logs and piles of debris. They measured 15"x 6½" x 4½".His print in this hard, damp sand was " to 1" deep where my print beside his was 1/8" deep.
— Taxidermist Bob Titmus writing to John Green, Nov. 7, 1958.

Dr. R. Maurice Tripp, geologist and geophysicist, has a cast of a footprint 17 inches long he made in the Bluff Creek area. Dr. Tripp"s engineering studies of the soil properties and depth of the foot print of which he made a cast show the weight of the owner of the print to be more than 800 pounds.
— *San Jose News*, 1958 or 1959.

17

Meet the Sasquatch

In December, 1979, I received a letter from Russell Gebhart of Lewiston, Idaho, which contained a copy of a report filed by an Idaho deputy sheriff, as follows:

Date: 11-11-79
Re: Sighting of an unknown type animal.
This is an officer's report by Thomas E. Dillon, 2007, Clearwater County Sheriff's Office, this is in reference of the sighting of an unknown type animal, approximately 1.2 miles over French Mountain Saddle, Clearwater County, Idaho, on the 25th day of October, 1979.

While on routine patrol of French Mountain Road, at approximately 0300 hours, this officer did observe, heading towards Bungalow, it was a foggy night, slight rain mist falling. While on routine patrol I did observe out of my peripheral vision something which I determined to be an animal. I backed up, took a handheld three hundred thousand candlepower spotlight, put it on this thing, at the first sighting it was approximately fifteen yards away and moving away from me, what I saw was a hairy animal covered in longish, matted unclean-looking brown hair, approximately the color of an elk's coat. This animal was moving away at a high rate of speed, it was approximately eight to eight and a half feet tall, approximately three to three and a half feet wide shoulders. I followed it for approximately eighty yards with my spotlight up a ravine, uphill, it covered this eighty yards in approximately four to five seconds. The animal was not moving in a jerky motion, but had a very fluid stride and approximately five to six foot span between his strides.

Returned to the scene the next day, did not find any tracks due to the rain, however I did find where it went through a patch of buck brush and did break the brush, no hair or other physical signs at the scene. I came to my realization as to the height from the comparison on the tree that it went by, where the animal's head was approximately two feet higher than mine.

As I said, I followed this animal for approximately eighty yards with my spotlight, it went up the ravine, uphill, covered this eighty yards in four to five seconds, moving at an extremely high rate of speed, going up the hill, it was running in a straddly type motion, throwing its legs to the sides, very rarely bending them. It had long arms, it was extremely powerfully built. The feeling I got from this animal was that it was extremely powerful, possibly dangerous. I did not exit my patrol vehicle at the time.

I believe that what I saw is what is called a Big Foot or Sasquatch.
End of report.

That is only one of many reports I have received from Russ Gebhart, a long-time associate in the sasquatch investigation, and only one of more than 2,000 reports in my files at time of writing. Because it is an official account of close, clear observations by a professional investigator, including action that would be beyond the physical capacity of a human, it ranks as one of the very best of all reports. It does not prove that the creature described by the deputy actually existed, but it demonstrates very clearly the one point that is absolutely certain about the "legendary" hairy giants of the Pacific Northwest. They are not just a story out of the past. The evidence of their existence, for whatever it is worth, is concentrated overwhelmingly in the present, right here, right now.

I call the giants "sasquatch" because that is what they are called by the Indians living in the area where I first learned about them. Up and down the coast they have different names, and most of those names are also of Indian origin. But that doesn't make the sasquatch an Indian legend—the Indians had names for all the other animals as well. The difference is that for the other animals the white man brought his own names with him. He has never named the sasquatch because no matter how often he sees them he refuses to accept that they are there.

Among the Indians there are other names with much wider distribution than the name "sasquatch", which seems to be limited to the southwest corner of British Columbia, and is an Anglicized word at that. However "sasquatch" has gained wide acceptance in the white community, rivalled only by the "Bigfoot" of northern California, because it was under that name that the creatures received their first press notices, back in the 1920's and 30's.

The man who introduced the sasquatch to the outside world was J.W. Burns, who was for many years a teacher at the Chehalis Indian Reserve, on the Harrison River near Harrison Hot Springs. His articles achieved wide circulation in papers and magazines in the United States and Canada. Like most native British Columbians I read some of his stories as a child. Mr. Burns got his stories from his Indian friends, and some of them smacked heavily of the supernatural. Thus he deserves not only the credit for being the first to introduce the sasquatch to the general public, but also the blame for fixing firmly to it the label of "Indian legend."

15-inch, barefoot tracks on a mountain road. Compare size with boot prints.

When I bought the weekly newspaper at Agassiz, B.C. in 1954 I knew that I was coming to the area occupied by Mr. Burns' sasquatch; in fact I once wrote an "April Fool" story for my paper about one of the hairy monsters kidnapping a beautiful guest from the Harrison Hot Springs Hotel. Of course I never took the subject seriously. Mr. Burns had left the area several years earlier, the Sasquatch Days once held at Harrison Hot Springs were a thing of the past and about the only reminder of the old stories was the Sasquatch Inn, a small hotel on the highway near the Harrison River. Today there is a Sasquatch Provincial Park beside Harrison Lake, but that came later.

That situation continued until 1956 when Rene Dahinden, a Swiss immigrant in his mid 20's, turned up at my office with the announcement that he intended to go into the mountains to look for the sasquatch. I tried to persuade him that the whole thing was just a tall story and referred him to a couple of veteran woodsmen whom I knew would assure him that there was no such thing to be found. He declined to be at all discouraged, but another man he had with him backed out and after a few days Rene went home.

In 1957 the subject came up again when the village council at Harrison Hot Springs was discussing what to do with about $600 available for a permanent project to mark British Columbia's 100th birthday. Not much of a permanent nature could be built for $600 even in those days, and one of the council members suggested using the money for a statue of a sasquatch. Next suggestion was that the money be used for a sasquatch hunt. That was greeted with enthusiasm, but since it was not the type of thing that was authorized by the provincial government for a project permission had to be sought from the B.C. Centennial Committee. It was, of course, a bid for publicity, and it was tremendously successful. Papers all over Canada played the story on the front page. There were numerous offers from would-be sasquatch hunters, even from young ladies prepared to act as bait. Typical of the light hearted enthusiasm was this column in the Vancouver *Daily Province* by humorist Eric Nicol:

Call me chicken if you like, but I can't go along with the idea of staking out a beautiful girl to lure a Sasquatch down from the Pitt Lake mountains.

As you've probably heard, a B.C. Centennial committee has set its heart on capturing one of the hairy giants and putting him on display during the '58 celebrations. The Sasquatch will represent British Colum-

SIGHTING AND TRACK REPORTS IN BRITISH COLUMBIA

The maps in this book are intended only to give a general impression of the number and location of sasquatch reports known to the author. Locations are plotted as accurately as possible, but in many cases no precise location is known and in others there are too many reports in the same locality to place them all correctly. Concentrations of reports along roads and water routes should not be taken to indicate that the sasquatch are concentrated there, but rather that there are people there to see them. In British Columbia in particular humans occupy only a very small percentage of the total area, while there are literally hundreds of thousands of square miles of mountains that are totally uninhabited and largely unvisited. To a lesser extent the same is true of the western parts of Washington, Oregon and Montana, and of northern California and Idaho.

MEET THE SASQUATCH

bia Before the Coming of the Safety Razor.

I don't mean that using a girl as bait wouldn't work to catch a Sasquatch. On the contrary, the Sasquatch are notoriously lickerish monsters. Their history is full of abductions of Indian maidens, who were returned after a year's trial and subsequently gave birth to a bouncing baby ghoul.

No doubt if you staked a beautiful girl out in those Harrison hills, and sprinkled a spoor of Yardley's Lavender you'd have Sasquatch queued up half way to Princeton.

But presumably there must also be a number of Centennial hunters on the scene, with nets and ropes and rifles. These are the ones I don't trust.

After all, that tangled jungle behind Harrison Lake can do strange things to a white man. And once the Sasquatch scented that there was small print involved in the transaction they would resort to their favorite trick of remaining invisible while throwing stones.

Let's say we have two hunters, Macomber and Flinch, stationed on a tree platform, watching the girl in her comfortably furnished snare in the clearing below. They have been on watch for a week. The weather has turned warm, and the girl reclines decoratively in a one-act play suit.

Stretching elaborately, Flinch gets up and prepares to descend the tree ladder.

"Where do you think you're going?" snaps Macomber.

"Time to water the bait," grins Flinch. "I'm off to the old water-hole."

"You gave her a drink only an hour ago." Macomber's eyes, reddened from six nights of watching the forest and slightly strabismed from also watching Flinch, glint dangerously. "And every time you come back from the water-hole you've got rye on your breath."

"Sulphur, old boy. This Harrison country is lousy with health-giving hot springs." Flinch's grinning, studded face disappears below the platform.

"Hold on, Flinch," Macomber raises himself on his elbows. "Something else I've noticed, every time you go to the water-hole the Sasquatch start piling rocks into this tree. One of them almost brained me this morning."

"Then keep your head down," yells Flinch, jumping to the ground.

Flinch disappears into the forest. A moment later a large stone crashes through the leaves of the tree, braining Macomber.

Wild and horrible laughter comes from the underbrush. As the girl scrambles to her feet, there is a quick movement behind her, and a hairy arm seizes her around the waist.

"Please, Mr. Flinch, not so tight," giggles the girl.

In an instant she has been whisked from the clearing. Silence. A figure lies still on the tree platform. And on the edge of the clearing a curiously twisted arm stretches out towards a brown bottle.

Well, as I said, I don't want to spoil anybody's fun. But I feel it's my duty to point out to the Harrison Centennial Committee that the Sasquatch plays rough, even by the standards of today's young women.

Instead of using girl-bait, therefore, I suggest they just watch for ads in the paper. They should be able to pick up a sasquatch at one of those monster sales. No?

Perhaps never before has a tourist resort achieved such publicity without actually doing anything. Rene Dahinden came back to lead the search. Newspaper and radio reporters flocked around, and a tide of delighted stories rolled around the world, touching such points as Sweden, India, New Zealand and places in between. No matter that the provincial Centennial Committee turned down the whole project and the hunt never took place. Public attention remained focussed on the sasquatch for months, with the news media digging up new angles to keep the fun going. In the process they discovered some stories that didn't fit the traditional pattern.

Mr. Burns' "Sasquatch" were basically giant Indians, with clothes, fire, weapons and the like. They were called hairy giants but most people just took that to mean long hair on their heads. William Roe, then living in Cloverdale, B.C., claimed to have seen something very different, a six-foot female weighing about 300 pounds that was covered from head to foot with dark brown hair less than an inch in length. This creature ate leaves by stripping them from a bush with its teeth. It wore no clothes.

Publication of his story brought him a letter from Albert Ostman, of Fort Langley, who claimed to have been carried off by one of the creatures, and to have spent several days with four of them in an alpine valley. He described in detail a mature male and female, a young male and an adolescent female. All were clothed only in short hair, and the big male was eight feet tall and very heavy.

Certainly these were not the sasquatch of the earlier newspaper stories and magazine articles. Nor were they characters from Indian legends. The men concerned were not Indians and they claimed to have been personally involved in the events they described.

The Ruby Creek Incident

The Roe and Ostman stories were most interesting, but they were only stories. There was no one else involved with whom they could be checked, and no records from the time and place where they were supposed to have happened. It was another incident altogether that forced me to begin taking the subject seriously—one in which there were plenty of supporting witnesses and also a certain amount of documentation.

The events had taken place some 13 years before, and had been brought to public attention, as was usual in those days, by Mr. Burns. The story, in the Vancouver *Province*, October 21, 1941, was located inconspicuously on page 12, under the heading:

HUGE BEAR TERRORIZES INDIANS

A child's scream, the uproar of dogs and a frightened woman's hurried glance led to tales among Ruby Creek Indians today of a huge hairy monster preying on their encampment.

It turned out to be a bear—but a huge one.

Rosie, small daughter of Mrs. George Chadwick, an Indian, was playing in her garden, half a mile east of Ruby Creek when she suddenly looked up to see the enormous beast approaching. She screamed for help.

Her mother rushed to her, got one glimpse of the monster, swept the child in her arms and dashed into the bush, where she remained for three hours before venturing home again.

On her return she found the racks of salted salmon scattered in every direction, but nothing else about the premises was touched.

In describing the animal, Mrs. Chadwick declared it was 10 feet tall, hairy, with a human face.

Little credence was given to the story until the beast returned. This time it left tracks revealing it to be one of the largest bears ever known in the vicinity. Its hind footmarks measured eight inches across and eighteen inches long. The span between the strides was five feet.

The Indians have requested the assistance of a game warden to destroy the monster.

That isn't a sasquatch story, of course. No mention of a sasquatch anywhere in it. In the first paragraph it started off in a promising way, talking of "a huge hairy monster," but in the next line "it turned out to be a bear."

Still, a little thought makes it look like a doubtful sort of bear story. "Ten feet tall ....with a human face" cannot successfully be applied to any bear. Then comes the part about tracks "revealing it to be one of the largest bears ever known in the vicinity." It had hind feet "eight inches across and eighteen inches long. The span between the strides was five feet." No bear on record could account for that set of statistics.

Presumably the editors responsible for the story were not students of nature.

In some ways that story is quite typical of newspaper accounts dealing the this subject. It uses words packed with excitement, playing up to the reader's taste for the exotic and mysterious yarn, but on the other hand it makes a point of emphasizing the commonplace explanation. It was only a bear after all.

Until the Harrison "sasquatch hunt" hit the headlines I had never heard of this incident, although Ruby Creek is only 12 miles up the Fraser River from Agassiz. But with all the publicity going on, the subject of sasquatches tended to come up in many conversations, and it did so when my wife and I were visiting Jack Kirkman, game guide at Harrison Hot Springs, and his Indian wife, Martha.

Martha Kirkman told us the story of the sasquatch at Ruby Creek as it had been told to her by her cousin Jeannie Chapman (not Chadwick) the woman who saw the creature. Mrs. Kirkman also said that when she was young there were places in the woods where the children were not allowed to go because the sasquatches were there. She did not say that she herself believed such creatures existed, but she did impress on us very strongly that Mrs. Chapman was serious in telling her story, and indeed had suffered a shock that changed her whole life.

On the same weekend Bill Rae, a printer who worked for me, was told the same story by Esse Tyfting, the head custodian of Agassiz High School, who had lived at Ruby Creek at that time. He had not seen the creature itself but was one of many local people who had studied the footprints that it left behind, and had found that the tracks confirmed Mrs. Chapman's account of the creature's movements.

Thoroughly intrigued, I went to see Mr. Tyfting, who repeated his story and drew an outline of a footprint for me on the floor of a room he was building. His story, and the size of the print he drew, were very impressive. He was a man whom I already knew, and whom I knew to have an excellent reputation in the community. When he said that he had actually seen those huge footprints I had no grounds to doubt him. He

THE RUBY CREEK INCIDENT

told me about other people who had been to look at the prints, and I was able to talk to several of them. Their recollections varied considerably, although all but one agreed that the prints could not have been made by any man or known animal. The lone exception insisted it must have been a bear, but he agreed with the others that it had walked on its hind legs and had stepped over a four-foot fence. I also went to see Mr. and Mrs. Chapman, talking to them on two occasions, and I visited their former home which had stood abandoned ever since the sasquatch came there.

Mrs. Chapman told me that one of her children had come to the house shouting about a "big cow coming out of the woods." She looked out the window and saw a man-like creature about eight feet tall and covered all over with fairly dark hair. It was walking across a field towards the house. She did not see its face from close up, but she was sure that it had a flat nose, not a snout like a bear.

Bears were very common around Ruby Creek at that time, and she was thoroughly familiar with their appearance.

Although terrified, Mrs. Chapman was still able to think clearly. She took the children and led them out the front door, keeping the house between her and the creature. They crossed a stretch of field and got down to the river, where a high cutbank shielded them from view. She did not know if the sasquatch saw them, but it did not try to follow. The tracks later showed that

the creature had circled the house and entered a shed where there was a barrel of salt salmon. He sampled this; there was some disagreement as to whether he had lifted and dumped it, but in any event there was torn fish scattered around. Then he went down to the river, perhaps to wash the salt out of his mouth, and returned to the mountainside.

I did not consider her story reliable as to detail, particularly as it was not entirely consistent, and I have since read accounts in which she is quoted as having said things which do not agree with some of the things she said to me.

I have noticed since that time that many people tend to reject an entire story if they can find fault with something in it—even a detail that has nothing whatever to do with the subject at hand. Several years as a reporter covering court cases have given me a more realistic view of the average person's ability to remember. No two witnesses, however impartial, ever have the same recollection of details of the same event, and it is rare for a witness who is testifying at any length to give precisely the same information at the trial that he gave at the preliminary hearing.

On the other hand most people do not lie very convincingly under questioning (some politicians excepted) and I was quite sure that Mrs. Chapman believed what she told me.

Later I talked to a son of the late Joe Dunn, a deputy sheriff from Whatcom County in Washington, who had investigated the

This is the house near Ruby Creek where the sasquatch came in 1941. Tracks led down from the mountains and across the field and circled behind the building.

THE RUBY CREEK INCIDENT

Ruby Creek incident at the time. Apparently he was already interested in the sasquatch as a result of experiences of his own. At his home I saw a report written by him that generally confirmed what I had learned myself, and was also able to copy a tracing of a footprint. By that time I had also talked to William Roe and Albert Ostman, and had heard two or three more reports of sightings that involved something more like an upright ape than the giant hairy Indians of the sasquatch stories.

Interviewing people had been part of my regular work for more than a decade, and aside from the basic improbability of the stories I could detect no indication that any of these people were not telling the truth, but I took the additional step of having some of them give sworn declarations that their stories were true. At the time I was under the impression that sworn stories would be taken more seriously by the scientists whom I hoped would take over the investigation, but that proved to be a mistake on my part.

In the case of Albert Ostman, as well as four of the witnesses from Ruby Creek, I even arranged for the local magistrate, a former trial lawyer of some reputation, to cross-examine them before taking their declarations. The following is a brief portion of my questioning of Esse Tyfting, recorded by the magistrate's secretary and later sworn to:

Q-Tell us what you saw.

A-Well it all started with Mrs. Chapman running down the track....crying "The sasquatch is after me!"

Q-What did you do?

A-I took the hand car....up to her place.

Q-And what did you find there?

A-The fish barrel had been turned over and there were fish all around the side of the house and we found prints going toward the river, leading from the potato patch to the edge of the C.P.R. fence and across the tracks and slough towards the mountain.

Q-How big were the tracks?

A-About 16 inches long, four inches at the heel and eight inches across the ball of the foot.

Q-Were there five toes?

A-Yes, but no claw marks....the stride between the prints was four feet between the heel and toe, all through the potato patch.

Q-Did you measure the footprints?

A-Yes.

Q-You, yourself?

A-Yes. I measured them and after a man

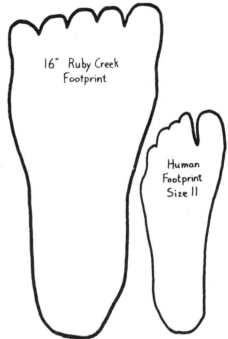

16" Ruby Creek Footprint

Human Footprint Size 11

came from across the line (Deputy Sheriff Dunn) we measured them again.

Q-And what did Mrs. Chapman say about the sasquatch?

A-She said he was a big hairy man.

Q-How did she describe the incident?

A-She said he looked through the window and she grabbed the children and ran down the tracks.

Q-And what condition was she in?

A-Scared to death.

Q-Was there anything about the C.P.R. fence that was particularly striking?

A-Well, we could see one footprint on this side (indicating) and another on this side (indicating).

Q-And how high was the fence?

A-Between four and five feet.

Q-The creature was able to step right over the fence?

A-Yes, not jump, just step.

Q-How deep were the footprints?

A-In the potato patch they were about two inches deep.

Q-And on that basis could you estimate how much this creature would weigh to make such a footprint?

A-I would say about 1,000 pounds—800 to 1,000 pounds to make a print that deep.

Q-Are you familiar with bear tracks?

A-Yes. I've seen enough of them at Ruby Creek.

Q-Could these have been bear tracks?

A-No, they certainly couldn't. Bear tracks wouldn't ever have the shape of a human heel. These looked like human feet.

Q-Any sign of the creature having walked on all fours?

A-No.

William Roe
and Albert Ostman

The reports of encounters with sasquatches that I have accumulated in more than 20 years of investigation are far too numerous to repeat in a book, even in summary, but among hundreds of eye-witness accounts now available, that of William Roe is still outstanding because of the variety of detail observed. His experience was most unusual in that he was able to watch at close range a sasquatch that did not know he was there, and then to backtrack it for a considerable distance. He is also one of the few people, all in the first few months of the investigation, who were asked to go to the trouble of having a declaration sworn attesting to the truth of his story, and he was the very first to describe a sasquatch as an apelike creature rather than a giant Indian.

As to Albert Ostman's story, it is unique and will probably remain so. In general outline it defies belief, and neither Rene Dahinden nor myself paid any attention to it until we were approached by an experienced radio reporter who had interviewed Mr. Ostman and was greatly impressed by him. When we talked to him ourselves we were also impressed, as was the magistrate who later interviewed him.

There is no way now to check the story, and considering that it was not written down until some 34 years after it happened it is certain to contain errors. But the description it gives of four sasquatches, male and female, old and young, are supported in general and in many details by information accumulated from scores of observers scattered over an area of hundreds of thousands of square miles. It is hard to see how Mr. Ostman, whose story was one of the very earliest, could have known so much without having had an opportunity for close observation of all the individuals he describes, and any sequence of events that could make such observation possible would be certain to sound unbelievable.

So that the reader may judge for himself, both stories are presented here in full with the supporting statutory declarations. William Roe had moved to Edmonton by the time I asked him to put the story in sworn form. Here is what he sent me:

Ever since I was a small boy back in the forest of Michigan I have studied the lives and habits of wild animals. Later, when I supported my family in northern Alberta by hunting and trapping I spent many hours just observing the wild things. They fasci-

nated me. But the most incredible experience I ever had with a wild creature occurred near a little town called Tete Jaune Cache, British Columbia, about eighty miles west of Jasper, Alberta.

I had been working on the highway near Tete Jaune Cache for about two years. In October, 1955, I decided to climb five miles up Mica Mountain to an old deserted mine, just for something to do. I came in sight of the mine about three o'clock in the afternoon after an easy climb. I had just come out of a patch of low brush into a clearing, when I saw what I thought was a grizzly bear, in the brush on the other side. I had shot a grizzly near that spot the year before. This one was only about 75 yards away, but I did not want to shoot it, for I had no way of getting it out. So I sat down on a small rock and watched, my rifle in my hands.

I could just see the top of the animal's head and the top of one shoulder. A moment later it raised up and stepped out into the opening. Then I saw that it was not a bear.

This, to the best of my recollection, is what the creature looked like and how it acted as it came across the clearing directly towards me. My first impression was of a huge man, about six feet tall, almost three feet wide and probably weighing somewhere near 300 pounds. It was covered from head to foot with dark brown, silver-tipped hair. But as it came closer I saw by its breasts that it was a female.

And yet, its torso was not curved like a female's. Its broad frame was straight from shoulder to hip. Its arms were much thicker than a man's arms, and longer, reaching almost to its knees. Its feet were broader proportionately than a man's, about five inches wide at the front and tapering to much thinner heels. When it walked it placed the heel of its foot down first, and I could see the grey-brown skin or hide on the soles of its feet.

It came to the edge of the bush I was hiding in, within twenty feet of me, and squatted down on its haunches. Reaching out its hands it pulled the branches of bushes toward it and stripped the leaves with its teeth. Its lips curled flexibly around the leaves as it ate. I was close enough to see that its teeth were white and even.

The shape of this creature's head somewhat resembled a negro's. The head was higher at the back than at the front. The nose was broad and flat. The lips and chin protruded farther than its nose. But the hair that covered it, leaving bare only the parts of the face around the mouth, nose and ears, made it resemble an animal as much as a human. None of its hair, even on the

WILLIAM ROE and ALBERT OSTMAN

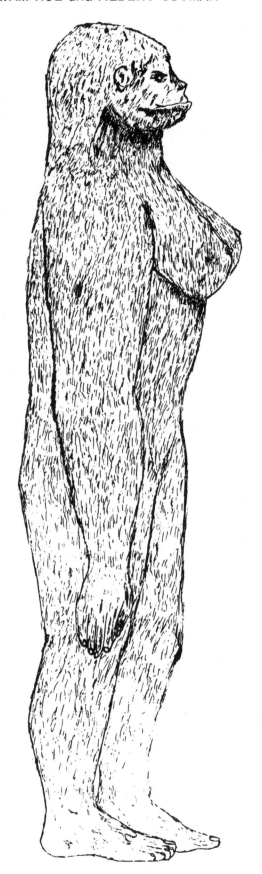

This drawing of the animal William Roe saw was done by his daughter, under his direction.

back of its head, was longer than an inch, and that on its face was much shorter. Its ears were shaped like a human's ears. But its eyes were small and black like a bear's. And its neck was unhuman. Thicker and shorter than any man's I had ever seen.

As I watched this creature, I wondered if some movie company was making a film at this place and that what I saw was an actor made up to look partly human and partly animal. But as I observed it more I decided it would be impossible to fake such a specimen. Anyway, I learned later that there was no such company near that area. Nor, in fact, did anyone live up Mica Mountain, according to the people who lived in Tete Jaune Cache.

Finally, the wild thing must have got my scent, for it looked directly at me through an opening in the brush. A look of amazement crossed its face. It looked so comical at the moment I had to grin. Still in a crouched position, it backed up three or four steps, then straightened up to its full height and started to walk rapidly back the way it had come. For a moment it watched me over its shoulder as it went, not exactly afraid, but as though it wanted no contact with anything strange.

The thought came to me that if I shot it, I would possibly have a specimen of great interest to scientists the world over. I had heard stories about the Sasquatch, the giant hairy Indians that live in the legends of British Columbia Indians, and also, many claim, are still in fact alive today. Maybe this was a Sasquatch, I told myself.

I levelled my rifle. The creature was still walking rapidly away, again turning its head to look in my direction. I lowered the rifle. Although I have called the creature "it," I felt now that it was a human being and I knew I would never forgive myself if I killed it.

Just as it came to the other patch of brush it threw back its head and made a peculiar noise that seemed to be half laugh and half language, and which I can only describe as a kind of whinny. Then it walked from the small brush into a stand of lodgepole pine.

I stepped out into the opening and looked across a small ridge just beyond the pine to see if I could see it again. It came out on the ridge a couple of hundred yards away from me, tipped its head back again, and again emitted the only sound I had heard it make, but what this half-laugh, half-language was meant to convey, I do not know. It disappeared then, and I never saw it again.

I wanted to find out if it lived on vegetation entirely or ate meat as well, so I went

WILLIAM ROE and ALBERT OSTMAN

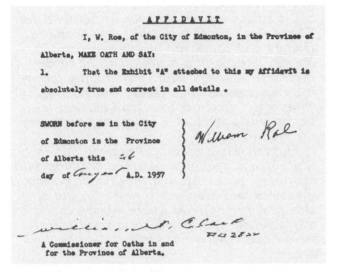

AFFIDAVIT

I, W. Roe, of the City of Edmonton, in the Province of Alberta, MAKE OATH AND SAY:

1.   That the Exhibit "A" attached to this my Affidavit is absolutely true and correct in all details.

SWORN before me in the City
of Edmonton in the Province
of Alberta this
day of   A.D. 1957

William Roe

William M. Clark

A Commissioner for Oaths in and
for the Province of Alberta.

down and looked for signs. I found it in five different places, and although I examined it thoroughly, could find no hair or shells of bugs or insects. So I believe it was strictly a vegetarian.

I found one place where it had slept for a couple of nights under a tree. Now, the nights were cool up the mountain, at this time of year especially, and yet it had not used a fire. I found no sign that it possessed even the simplest of tools. Nor a single companion while in this place.

Whether this creature was a Sasquatch I do not know. It will always remain a mystery to me, unless another one is found.

I hereby declare the above statement to be in every part true, to the best of my powers of observation and recollection.

WILLIAM ROE

Sworn before William Clark, a Commissioner for Oaths in and for the Province of Alberta.

In addition to the information in this sworn statement, Mr. Roe made the following remarks regarding the sasquatch in a letter:

The nails were not like a bear's, but short and heavy like a man's finger nails are. Its eyes were not light and large but small and black like a bear's. You couldn't see any knotted, corded muscles. This animal seemed almost round. It was as deep through as it was wide, and I believe if this animal should have been seven feet tall, it would have weighed close to 500 pounds.

We have got to get away from the idea of comparing it to a human being as we know them.

I never did meet Mr. Roe and I knew very little about him, but in 1969 on a trip across Canada I met two zoologists in different

cities who had corresponded with him concerning his observations of buffalo. They both considered him to be a well-qualified and reliable student of wildlife.

Albert Ostman had written his story before I met him. When he was asked to recall all he could of his encounter with the sasquatch back in 1924 he went about it by gathering whatever he could locate   from that other period in his life, including among other things a shopping list for one of his prospecting trips. Then he set about rebuilding the experience in detail, including his own actions prior to and following the actual encounter, in an attempt to re-enter, as much as possible, the scene of events that took place more than 30 years ago.

When he was later asked if he would swear to the accuracy of the account he made it clear that he could do so only as to the main elements of the story, not the surrounding detail. Here is the story he wrote:

I have always followed logging and construction work. This time I had worked for over one year on a construction job and thought a good vacation was in order. B.C. is famous for lost gold mines. One is supposed to be at the head of Toba Inlet—why not look for this mine and have a vacation at the same time? I took the Union Steamship boat to Lund, B.C. From there I hired an old Indian to take me to the head of Toba Inlet.

This old Indian was a very talkative old gentleman. He told me stories about gold brought out by a white man from this lost mine. This white man was a very heavy drinker—spent his money freely in saloons. But he had no trouble getting more money. He would be away a few days, then come back with a bag of gold. But one time he went to his mine and never came back. Some people said a Sasquatch had killed him.

At that time I had never heard of Sasquatch. So I asked what kind of an animal he called a Sasquatch. The Indian said, "They have hair all over their bodies, but they are not animals. They are people. Big people living in the mountains. My uncle saw the tracks of one that were two feet long. One old Indian saw one over eight feet tall."

I told the Indian I didn't believe in their old fables about mountain giants. It might have been some thousands of years ago, but not nowadays.

The Indian said, "There may not be many, but they still exist."

We arrived at the head of the inlet at 4:00 p.m. I made camp at the mouth of a creek. The Indian was in no hurry, he had to wait

27

WILLIAM ROE and ALBERT OSTMAN

for the high tide to go back. That would be about 7:00 p.m. I tried to catch some trout in the creek, but no luck. The Indian had supper with me, and I told him to look out for me in about three weeks. I would be camping at the same spot when I came back. He promised to tell his friend to look out for me too.

I spent most of the forenoon looking for a trail but found none, except for a hog-back running down to within about a hundred feet of the beach. So I swamped out a trail from there, got back to my camp about 3:00 p.m. that afternoon and made up my pack to be ready in the morning. My equipment consisted of one 30-30 Winchester rifle, I had a special home-made prospecting pick, axe on one end, and pick on the other. I had a leather case for this pick which fastened to my belt, also my sheath knife.

Next morning I took my rifle with me, but left my equipment at the camp. I decided to look around for some deer trail to lead me up into the mountains. On the way up the inlet I had seen a pass in the mountainside that I wanted to go through, to see what was on the other side.

The storekeeper at Lund was co-operative. He gave me some cans for my sugar, salt and matches to keep them dry. My grub consisted mostly of canned stuff, except for a side of bacon, a bag of beans, four pounds of prunes and six packets of macaroni, three pounds of pancake flour, cheese, and six packets Rye King hard tack, three rolls of snuff, one quart sealer of butter and two one-pound cans of milk. I had two boxes of shells for my rifle.

The storekeeper gave me a biscuit tin, I put a few things in that and cached it under a windfall, so I would have it when I came back here waiting for a boat to bring me out. My sleeping bag I rolled up and tied on top of my pack sack—together with my ground sheet, small frying pan, and one aluminum pot that held about a gallon. As my canned food was used, I would get plenty of empty cans to cook with.

The following morning I had an early breakfast, made up my pack, and started up this hogback. My pack must have been at least eighty pounds, besides my rifle. After one hour, I had to rest. I kept resting and climbing all that morning. About 2:00 p.m. I came to a flat place below a rock bluff. There was a bunch of willow in one place. I made a wooden spade and started digging for water. About a foot down I got seepings of water, so I decided to camp here for the

night, and scout around for the best way to get on from here.

I must have been up near a thousand feet. There was a most beautiful view over the islands and the Strait—tugboats with log booms, and fishing boats going in all directions. A lovely spot. I spent the following day prospecting round. But no sign of minerals. I found a deer trail leading towards this pass that I had seen on my way up the inlet.

The following morning I started out early, while it was cool. It was steep climbing with my heavy pack. After a three hours climb, I was tired and stopped to rest. On the other side of a ravine from where I was resting was a yellow spot below some small trees. I moved over there and started digging for water.

I found a small spring and made a small trough from cedar bark and got a small amount of water, had my lunch and rested here till evening. This was not a good camping site, and I wanted to get over the pass. I saved all the water I got from this spring, as I might not find water on the other side of this pass. However, I made it over the pass late that night.

Now I had downhill and good going, but I was hungry and tired, so I camped at the first bunch of trees I came to. I had about a gallon of water so I was good for one day. Of course, I could see rough country ahead of me, and I was trying to size up the terrain—what direction I would take from here. Towards the west would lead to low land and some other inlet, so I decided to go in a northeast direction, but I had to find a good way to get down.

I left my pack and went east along a ledge but came to an abrupt end—was two or three hundred feet straight down. I came back, found a place only about 50 feet down to a ledge that looked like good going for as far as I could see. I got down on this ledge all right and had good going and slight down hill all day—I must have made 10 miles when I came to a small spring and a big black hemlock tree.

This was a lovely campsite. I spent two days here just resting and prospecting. There were some minerals but nothing interesting. The first night here I shot a small deer (buck) so I had plenty of good meat, and good water. The weather was very hot in the daytime, so I was in no hurry, as I had plenty of meat. When I finally left this camp, I got into plenty of trouble. First I got into a box canyon, and had to come back almost where I started this morning, when I found a deer trail down to another ledge, and had about two miles of good going. Then

WILLIAM ROE and ALBERT OSTMAN

I came to another canyon, and on the other side was a yellow patch of grass that meant water. I made it down into this canyon, and up on the other side, but it was tough climbing. I was tired and when I finally got there I dug a pit for water and got plenty for my needs. I only stayed here one night, it was not a good camping site. Next day I had hard going. I made it over a well timbered ridge into another canyon. This canyon was not so steep on the west side, but the east side was almost plumb. I would have to go down hill to find a way out. I was now well below the timber line.

I found a fair campsite that night, but moved on the next morning. It was a very hot day, not a breath of wind.

Late that day I found an exceptionally good campsite. It was two good-sized cypress trees growing close together and near a rock wall with a nice spring just below these trees. I intended to make this my permanent camp. I cut lots of brush for my bed between these trees. I rigged up a pole from this rock wall to hang my packsack on, and I arranged some flat rocks for my fireplace for cooking. I had a really classy setup. I shot a grouse just before I came to this place. Too late to roast that tonight—I would do that tomorrow.

And that's when things began to happen.

I am a heavy sleeper, not much disturbs me after I go to sleep, especially on a good bed like I had now.

Next morning I noticed things had been disturbed during the night. But nothing missing that I could see. I roasted my grouse on a stick for my breakfast—about 9:00 a.m. I started out prospecting. I always carried my rifle with me. Your rifle is the most important part of your equipment.

I started out in a southwest direction below the way I had come in the night before. There were some signs (minerals) but nothing important. I shot a squirrel in the afternoon, and got back to camp at 7:00 p.m. I fried the squirrel on a stick, opened a can of peas and carrots for my supper, and gathered up dry branches from trees. There are always dead branches of fir and hemlock under trees, near the ground. They make good fuel and good heat.

That night I filled up the magazine of my rifle. I still had one full box of 20 shells in my pack, besides a full magazine and six shells in my coat pocket. That night I laid my rifle under the edge of my sleeping bag. I thought a porcupine had visited me the night before and porkies like leather, so I put my shoes in the bottom of my sleeping bag.

Next morning my pack sack had been emptied out. Someone had turned the sack upside down. It was still hanging on the pole from the shoulder straps as I had hung it up. Then I noticed one half-pound package of prunes was missing. Also my pancake flour was missing, but my salt bag was not touched. Porkies always look for salt, so I decided it must be something else than porkies. I looked for tracks but found none. I did not think it was a bear, they always tear up and make a mess of things. I kept close to camp those days in case this visitor would come back.

I climbed up on a big rock where I had a good view of the camp, but nothing showed up. I was hoping it would be a porky so I could get a good porky stew. These visits had now been going on for three nights.

I intended to make a new campsite the following day, but I hated to leave this place. I had fixed it up so nicely, and these two cypress trees were bushy. It would have to be a heavy rain before I would get wet, and I had good spring water and that is hard to find.

This night it was cloudy and looked like it might rain. I took special notice of how everything was arranged. I closed my pack sack, I did not undress, I only took off my shoes, put them in the bottom of my sleeping bag. I drove my prospecting pick into one of the cypress trees so I could reach it from my bed. I also put the rifle alongside me, inside my sleeping bag. I fully intended to stay awake all night to find out who my visitor was, but I must have fallen asleep.

I was awakened by something picking me up. I was half asleep and at first I did not remember where I was. As I began to get my wits together, I remembered I was on this prospecting trip, and in my sleeping bag.

My first thought was—it must be a snow slide, but there was no snow around my camp. Then it felt like I was tossed on horseback, but I could feel whoever it was, was walking.

I tried to reason out what kind of animal this could be, I tried to get at my sheath knife, and cut my way out, but I was in an almost sitting position, and the knife was under me. I could not get hold of it, but the rifle was in front of me, I had a good hold of that, and had no intention to let go of it. At times I could feel my packsack touching me, and could feel the cans in the sack touching my back.

After what seemed like an hour, I could feel we were going up a steep hill. I could feel myself rise for every step. What was carrying me was breathing hard and sometimes gave a slight cough. Now, I knew this

WILLIAM ROE and ALBERT OSTMAN

CANADA   )
)   IN THE MATTER OF "THE SASQUATCH"
Province of British Columbia   )
)
            TO WIT:)

I, Albert Ostman, of Langley Municipality in the Province of British
Columbia, retired, do solemnly declare:

   That the attached article, signed by me and marked Exhibit "A"
   is a true copy of the events which happened as set forth therein.

   AND I make this solemn Declaration conscientiously believing it
   to be true, and knowing that it is of the same force and effect
   as if made under oath and by virtue of The Canada Evidence Act.

DECLARED before me at Langley   )
Municipality in the Province   )
of British Columbia, the   )
Twentieth day of August, A.D.   )   _Albert Ostman_
1957   )
)
    A. M. Naismith   )
)
A Justice of the Peace in and
for the Province of British
Columbia.

must be one of the mountain Sasquatch
giants the Indian told me about.

I was in a very uncomfortable position,
unable to move. I was sitting on my feet,
and one of the boots in the bottom of the
bag was crossways with the hobnail sole up
across my foot. It hurt me terribly, but I
could not move.

It was very hot inside. It was lucky for
me this fellow's hand was not big enough to
close up the whole bag when he picked me
up—there was a small opening at the top.
Otherwise I would have choked to death.

Now he was going downhill, I could feel
myself touching the ground at times and at
one time he dragged me behind him and I
could feel he was below me. Then he seemed
to get on level ground and was going at a
trot for a long time. By this time, I had
cramps in my legs, the pain was terrible. I
was wishing he would get to his destination
soon. I could not stand this type of trans-
portation much longer.

Now he was going uphill again. It did not
hurt me so bad. I tried to estimate distance
and directions. As near as I could guess we
were about three hours travelling. I had no
idea when he started as I was asleep when
he picked me up.

Finally he stopped and let me down. Then
he dropped my packsack, I could hear the
cans rattle. Then I heard chatter—some kind
of talk I did not understand. The ground
was sloping so when he let go of my sleeping
bag, I rolled over head first downhill. I got
my head out, and got some air. I tried to
straighten my legs and crawl out, but my
legs were numb.

It was still dark, I could not see what my
captors looked like. I tried to massage my
legs to get some life in them, and get my
shoes on. I could hear now it was at least
four of them. They were standing around me
and continuously chattering. I had never

heard of Sasquatch before the Indian told me
about them. But I knew I was right among them.

But how to get away from them, that was
another question. I got to see the outline of
them now, as it began to get lighter....

I now had circulation im my legs, but my
left foot was very sore on top where it had
been resting on my hobnail boots. I got my
boots out from the sleeping bag and tried to
stand up. I was wobbly on my feet but had
a good hold of my rifle.

I asked, "What you fellows want with me?"
Only some more chatter.

It was getting lighter now, and I could see
them quite clearly. I could make out forms of
four people. Two big ones and two little ones.
They were all covered with hair and no
clothes on at all.

I could now make out mountains all around
me. I looked at my watch. It was 4:25 a.m.
It was getting lighter now and I could see
the people clearly.

They look like a family, old man, old lady
and two young ones, a boy and a girl. The
boy and the girl seem to be scared of me.
The old lady did not seem too pleased about
what the old man dragged home. But the old
man was waving his arms and telling them all
what he had in mind. They all left me then.

I had my compass and my prospecting
glass on strings around my neck. The compass in my lefthand shirt pocket and my glass
in my righthand pocket. I tried to reason
our location, and where I was. I could see
now that I was in a small valley or basin
about eight or ten acres, surrounded by
high mountains, on the southeast side there
was a V-shaped opening about eight feet
wide at the bottom and about twenty feet
wide at the highest point—that must be the
way I came in. But how will I get out? The
old man was now sitting near this opening.

I moved my belongings up close to the
west wall. There were two small cypress
trees there, and this will do for a shelter
for the time being. Until I find out what
these people want with me, and how to get
away from here. I emptied out my packsack
to see what I had left in the line of food. All
my canned meat and vegetables were intact
and I had one can of coffee. Also three small
cans of milk—two packages of Rye King hard
tack and my butter sealer half full of butter.
But my prunes and macaroni were missing.
Also my full box of shells for my rifle. I only
had six shells beside what I had in the magazine of my rifle. I had my sheath knife but
my prospecting pick was missing and my can
of matches. I only had my safety box full
and that held about a dozen matches. That
did not worry me—I can always start a fire
with my prospecting glass when the sun is

WILLIAM ROE and ALBERT OSTMAN

shining, if I got dry wood. I wanted hot coffee, but I had no wood, also nothing around here that looked like wood. I had a good look over the valley from where I was —but the boy and the girl were always watching me from behind some juniper bush. I decided there must be some water around here. The ground was leaning towards the opening in the wall. There must be water at the upper end of this valley, there is green grass and moss along the bottom.

All my utensils were left behind. I opened my coffee tin and emptied the coffee in a dishtowel and tied it with the metal strip from the can. I took my rifle and the can and went looking for water. Right at the head under a cliff there was a lovely spring that disappeared underground. I got a drink, and a full can of water. When I got back the young boy was looking over my belongings, but did not touch anything. On my way back I noticed where these people were sleeping. On the east side wall of this valley was a shelf in the mountain side, with overhanging rock, looking something like a big under-cut in a big tree about 10 feet deep and 30 feet wide. The floor was covered with lots of dry moss, and they had some kind of blankets woven of narrow strips of cedar bark, packed with dry moss. They looked very practical and warm—with no need of washing.

The first day not much happened. I had to eat my food cold. The young fellow was coming nearer me, and seemed curious about me. My one snuff box was empty, so I rolled it towards him. When he saw it coming, he sprang up quick as a cat, and grabbed it. He went over to his sister and showed her. They found out how to open and close it —they spent a long time playing with it—then he trotted over to the old man and showed him. They had a long chatter.

Next morning, I made up my mind to leave this place—if I had to shoot my way out. I could not stay much longer, I had only enough grub to last me till I got back to Toba Inlet. I did not know the direction but I would go down hill and I would come out near civilization some place. I rolled up my sleeping bag, put that inside my pack sack —packed the few cans I had—swung the sack on my back, injected a shell in the barrel of my rifle and started for the opening in the wall. The old man got up, held up his hands as though he would push me back.

I pointed to the opening. I wanted to go out. But he stood there pushing towards me —and said something that sounded like "Soka soka." I backed up to about sixty feet. I did not want to be too close, I thought, if I had to shoot my way out. A 30-30 might not

have much effect on this fellow, it might make him mad. I only had six shells so I decided to wait. There must be a better way than killing him, in order to get out from here. I went back to my campsite to figure out some other way to get out.

If I could make friends with the young fellow or the girl, they might help me. If I only could talk to them. Then I thought of a fellow who saved himself from a mad bull by blinding him with snuff in his eyes. But how will I get near enough to this fellow to put the snuff in his eyes? So I decided next time I give the young fellow my snuff box to leave a few grains of snuff in it. He might give the old man a taste of it.

But the question is, in what direction will I go, if I should get out? I must have been near 25 miles northeast of Toba Inlet when I was kidnapped. This fellow must have travelled at least 25 miles in the three hours he carried me. If he went west we would be near salt water—same thing if he went south —therefore he must have gone northeast. If I then keep going south and over two mountains, I must hit salt water someplace between Lund and Vancouver.

The following day I did not see the old lady till about 4:00 p.m. She came home with her arms full of grass and twigs of all kinds from spruce and hemlock as well as some kind of nuts that grow in the ground. I have seen lots of them on Vancouver Island. The young fellow went up the mountain to the east every day, he could climb better than a mountain goat. He picked some kind of grass with long sweet roots. He gave me some one day—they tasted very sweet. I gave him another snuff box with about a teaspoon of snuff in it. He tasted it, then went to the old man--he licked it with his tongue. They had a long chat. I made a dipper from a milk can. I made many dippers—you cut two slits near the top of any can—then cut a limb from any small tree—cut down back of the limb—down the stem of the tree—then taper the part you cut from the stem. Then cut a hole in the tapered part, slide the tapered part into the slit you made in the can, and you have a good handle on your can. I threw one over to the young fellow that was playing near my camp, he picked it up and looked at it then he went to the old man and showed it to him. They had a long chatter. Then he came to me, pointed at the dipper then at his sister. I could see that he wanted one for her too. I had other peas and carrots, so I made one for his sister. He was standing only eight feet away from me. When I had made the dipper, I dipped it in water and drank from it, he was very pleased, almost smiled at me. Then I took a chew of snuff,

31

WILLIAM ROE and ALBERT OSTMAN

smacked my lips, said that's good.

The young fellow pointed to the old man, said something that sounded like "Oook." I got the idea that the old man liked snuff, and the young fellow wanted a box for the old man. I shook my head. I motioned with my hands for the old man to come to me. I do not think the young fellow understood what I meant. He went to his sister and gave her the dipper I made for her. They did not come near me again that day. I had now been here six days, but I was sure I was making progress. If only I could get the old man to come over to me, get him to eat a full box of snuff that would kill him for sure, and that way kill himself, I wouldn't be guilty of murder.

The old lady was a meek old thing. The young fellow was by this time quite friendly. The girl would not hurt anybody. Her chest was flat like a boy—no development like young ladies. I am sure if I could get the old man out of the way, I could easily have brought this girl out with me to civilization. But what good would that have been? I would have to keep her in a cage for public display. I don't think we have any right to force our way of life on other people, and I don't think they would like it. (The noise and racket in a modern city they would not like any more than I do.)

The young fellow might have been between 11–18 years old about seven feet tall and might weigh about 300 lbs. His chest would be about 50–55 inches, his waist about 36–38 inches. He had wide jaws, narrow forehead that slanted upward round at the back about four or five inches higher than the forehead. The hair on their heads was about six inches long. The hair on the rest of their body was short and thick in places. The women's hair was a bit longer on their heads and the hair on their forehead had an upward turn like some women have—they call it bangs, among women's hair-do's. Nowadays the old lady could have been anything between 40–70 years old. She was over seven feet tall. She would be about 500–600 pounds

She had very wide hips, and a gooselike walk. She was not built for beauty or speed. Some of those lovable brassieres and uplifts would have been a great improvement on her looks and her figure. The man's eyeteeth were longer than the rest of the teeth, but not long enough to be called tusks. The old man must have been near eight feet tall. Big barrel chest and big hump on his back —powerful shoulders, his biceps on upper arm were enormous and tapered down to his elbows. His forearms were longer than common people have, but well proportioned. His

hands were wide, the palm was long and broad and hollow like a scoop. His fingers were short in proportion to the rest of the hand. His fingernails were like chisels. The only place they had no hair was inside their hands and the soles of their feet and upper part of the nose and eyelids. I never did see their ears, they were covered with hair hanging over them.

If the old man were to wear a collar it would have to be at least 30 inches. I have no idea what size shoes they would need. I was watching the young fellow's foot one day when he was sitting down. The soles of his feet seemed to be padded like a dog's foot, and the big toes was longer than the rest and very strong. In mountain climbing all he needed was footing for his big toe. They were very agile. To sit down they turned their knees out and came straight down. To rise they came straight up without help of their hands and arms. I don't think this valley was their permanent home. I think they move from place to place, as food is available in different localities. They might eat meat, but I never saw them eat meat, or do any cooking.

I think this was probably a stopover place and the plants with sweet roots on the mountainside might have been in season this time of the year. They seem to be most interested in them. The roots have a very sweet and satisfying taste. They always seem to do everything for a reason, wasted no time on anything they did not need. When they were not looking for food, the old man and the old lady were resting, but the boy and the girl were always climbing something or some other exercise. His favorite position was to take hold of his feet with his hands and balance on his rump, then bounce forward. The idea seems to be to see how far he could go without his feet or hands touching the ground. Sometimes he made 20 feet.

But what do they want with me? They must understand that I cannot stay here indefinitely. I will soon run out of grub, and so far I have seen no deer or other game. I will soon have to make a break for freedom. Not that I was mistreated in any way. One consolation was that the old man was coming closer each day, and was very interested in my snuff. Watching me when I take a pinch of snuff. He seems to think it it useless to only put it inside my lips. One morning after I had my breakfast both the old man and the boy came and sat down only ten feet away from me. This morning I made coffee. I had saved up all dry branches I found and I had some dry moss and I used all the labels from cans to start a fire.

I got my coffee pot boiling and it was

WILLIAM ROE and ALBERT OSTMAN

Albert Ostman

strong coffee too, and the aroma from boiling coffee was what brought them over. I was sitting eating hard-tack with plenty of butter on, and sipping coffee. And it sure tasted good. I was smacking my lips pretending it was better than it really was. I set the can down that was about half full. I intended to warm it up later. I pulled out a full box of snuff, took a big chew. Before I had time to close the box the old man reached for it. I was afraid he would waste it, and only had two more boxes. So I held on to the box intending him to take a pinch like I had just done. Instead he grabbed the box and emptied it in his mouth. Swallowed it in one gulp. Then he licked the box inside with his tongue.

After a few minutes his eyes began to roll over in his head, he was looking straight up. I could see he was sick. Then he grabbed my coffee can that was quite cold by this time, he emptied that in his mouth, grounds and all. That did no good. He stuck his head between his legs and rolled forwards a few times away from me. Then he began to squeal like a stuck pig. I grabbed my rifle. I said to myself, "This is it. If he comes for me I will shoot him plumb between his eyes." But he started for the spring, he wanted water. I packed my sleeping bag in the pack sack with the few cans I had left. The young fellow ran over to his mother. Then she

began to squeal. I started for the opening in the wall— and I just made it. The old lady was right behind me. I fired one shot at the rock over her head.

I guess she had never seen a rifle fired before. She turned and ran inside the wall. I injected another shell in the barrel of my rifle and started downhill, looking back over my shoulder every so often to see if they were coming. I was in a canyon, and good travelling and I made fast time. Must have made three miles in some world record time. I came to a turn in the canyon and I had the sun on my left, that meant I was going south and the canyon turned west. I decided to climb the ridge ahead of me. I knew I must have two mountain ridges between me and salt water and by climbing this ridge I would have a good view of this canyon, so I could see if the Sasquatch were coming after me. I had a light pack and was making good time up this hill. I stopped after to look back to where I came from, but nobody followed me. As I came over the ridge I could see Mount Baker. Then I knew I was going in the right direction.

I was hungry and tired. I opened my pack sack to see what I had to eat. I decided to rest here for a while. I had a good view of the mountainside, and if the old man was coming I had the advantage because I was above him. To get me he would have to come up a steep hill. And that might not be so easy after stopping a few 30-30 bullets. I had made up my mind this was my last chance, and this would be a fight to the finish. I ate some hard tack and I opened my last can of corned beef. I had no butter, I forgot to pick up my butter sealer I had buried near my camp to keep it cold. I did not dare to make a fire. I rested here for two hours. It was 3:00 p.m. when I started down the mountain side. It was nice going, not too steep, and not too much underbrush.

When I got near the bottom I shot a big blue grouse. She was sitting on a windfall, looking right at me, only a hundred feet away. I shot her neck right off.

I made it down to the creek at the bottom of this canyon. I felt I was safe now. I made a fire between two big boulders, roasted the grouse, made some coffee and opened my can of milk. My first good meal for days. I spread out my sleeping bag under a big spruce tree and went to sleep. Next morning when I woke up, I was feeling terrible. My feet were sore from dirty socks. My legs were sore, my stomach was upset from that grouse that I ate the night before. I was not sure I was going to make it up that mountain. It was a cloudy day, no sun, but after some coffee and hard tack I felt a bit better. I

33

WILLIAM ROE and ALBERT OSTMAN

started up the mountainside but had no energy. I only wanted to rest. My legs were shaking. I had to rest every hundred feet. I finally made the top, but it took me six hours to get there. It was cloudy, visibility about a mile.

I knew I had to go down hill. After about two hours I got down to the heavy timber and sat down to rest. I could hear a motor running hard at times, then stop. I listened to this for a while and decided the sound was a gas donkey. Someone was logging in the neighborhood. I made for this sound, for if only I can get to that donkey, I will be safe. After a while I hear someone holler "Timber" and a tree go down. Now I knew I was safe. When I came up to the fellows, I guess I was a sorry sight. I hadn't had a shave since I left Toba Inlet, and no good wash for days. When I came up out of the bushes, they kept staring at me. I asked where the place was and how far to the nearest town. The men said, "You look like a wild man, where did you come from?"

I told them I was a prospector and was lost. I had not had much to eat the last few weeks. I got sick from eating a grouse last night, and I am all in. The bucker called to his partner, "Pete, come over here a minute." Pete came over and looked at me and said this man is sick. We had better help him down to the landing, put him on a logging truck and send him down to the beach. I did not like to tell them I had been kidnapped by a Sasquatch, as if I had told them, they would probably have said, he is crazy too. They were very helpful and they talked to the truck driver to give me a ride down to the beach. Pete helped me up into the truck cab, and said the First Aid man will fix you up at the camp. The first aid man brought me to the cook and asked "Have you a bowl of soup for this man?" The cook came and looked me over. He asked, "When did you eat last, and where did you come from." I told him I had been lost in the wood. I ate a grouse last night and it made me sick.

After the cook had given me a first class meal, the first aid man took me to the first aid house. I asked "Can you get me a clean suit of underwear and a pair of socks? I would like a bath, too." He said, "Sure thing, you take a rest and I will fix all that. I'll arrange for you to go down to Sechelt when the timekeeper goes down for mail." After a session in the bathroom the first aid man gave me a shave and a hair trim, and I was back to my normal self. The Bull of the Woods told me I was welcome to stay for a day and rest up if I liked. I told him I accepted his hospitality as I was not feeling

any too good yet. I told about my prospecting, but nothing about being kidnapped by a sasquatch.

The following day I went down from this camp on the Salmon Arm Branch of Sechelt Inlet. From there I got the Union Boat back to Vancouver. That was my last prospecting trip, and my only experience with what is known as Sasquatches. I know that in 1924 there were four sasquatches living, it might be only two now. The old man and the old lady might be dead by this time.

That is the story. If someone told it now it would probably be laughed off, even by sasquatch enthusiasts, because detailed information is readily available in print and several people have made up slightly similar accounts of adventures with the hairy giants. For Albert Ostman there was no pattern to follow. Some of the things he said of the sasquatch have not been confirmed by the hundreds of later reports. No one else, for instance, has described anything like bark and moss blankets. But his descriptions of the creatures themselves, which were at variance with the common impression at that time, have been confirmed over and over again.

The only other information that has come to my attention which appears to have a bearing on the Ostman story was a comment by an old friend of mine that he had first heard of the sasquatch in the early 1930's from a trapper at Toba Inlet who said he knew a young Swede who had been carried off by one.

OFFICE OF
STIPENDIARY MAGISTRATE
POLICE MAGISTRATE
JUDGE OF JUVENILE COURT
JUSTICE OF THE PEACE
CORONER

PROVINCE OF
BRITISH COLUMBIA

P. O. BOX 52
HARRISON HOT SPRINGS, B.C.

August 20, 1957.

TO WHOM IT MAY CONCERN:

        THIS IS TO CERTIFY that on Tuesday, August 20, 1957, I accompanied Mr. John Green, Editor of the Agassiz-Harrison Advance to the home of Mr. Albert OSTMAN, located near Fort Langley, B.C. Mr. Green had visited Mr. Ostman on a previous occasion and obtained from him a story pertaining to "The Sasquatch".

        I found Mr. Ostman to be a man of sixty-four years of age; in full possession of his mental faculties; of pleasant manner, and with a good sense of humor.

        I questioned Mr. Ostman thoroughly in reference to the story given Mr. Green. I cross-examined him and used every means to endeavor to find a flaw in either his personality or his story, but could find neither.

        I certainly left with the impression that Mr. Ostman believes in his story himself and considers he is telling the truth. My examination and cross-examination failed to bring out any evidence to the contrary.

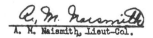

A. M. Naismith, Lieut-Col.

A Sasquatch Caught?

When a newspaperman starts to look into something, one of his first steps is to see what, if anything, is in the files. For the sasquatch Rene and I found that the best file was in the British Columbia Archives, at the Legislative Buildings in Victoria. Here we learned that there was a certain amount of material earlier than any of the Burns stories.

Bruce McKelvie, a newspaperman and historian, had done some research into two reports of sightings of hairy humans on Vancouver Island, and there were references to incidents in Washington, including the famous encounter at Ape Canyon on Mount St. Helens.

More important, there was a reference to a newspaper story telling of the capture of a small sasquatch during the construction of the railroad in the Fraser Canyon in the 1880's. Printed in the July 4, 1884 edition of the Victoria *Daily Colonist,* the story went as follows:

WHAT IS IT?

—

A STRANGE CREATURE CAPTURED ABOVE YALE

—

A British Columbia Gorilla

(Correspondence of The Colonist)

Yale, B.C., July 3rd, 1882.(stet)

In the immediate vicinity of No. 4 tunnel situated some twenty miles above this village, are bluffs of rock which have hitherto been unsurmountable, but on Monday morning last were successfully scaled by Mr. Onderdonk's employees on the regular train from Lytton. Assisted by Mr. Costerton, the British Columbia Express Company's messenger, and a number of gentlemen from Lytton and points east of that place, who, after considerable trouble and perilous climbing, succeeded in capturing a creature which may truly be called half man and half beast. "Jacko," as the creature has been called by his capturers, is something of the gorilla type standing about four feet seven inches in height and weighing 127 pounds. He has long, black, strong hair and resembles a human being with one exception, his entire body, excepting his hands, (or paws) and feet are covered with glossy hair about one inch long. His fore arm is much longer than a man's fore arm, and he possesses extraordinary strength, as he will take hold of a stick and break it by wrenching or twisting it, which no man living could break in the

same way. Since his capture he is very reticent, only occasionally uttering a noise which is half bark and half growl. He is, however, becoming daily more attached to his keeper, Mr. George Tilbury, of this place, who proposes shortly starting for London, England, to exhibit him. His favorite food so far is berries, and he drinks fresh milk with evident relish. By advice of Dr. Hannington raw meats have been withheld from Jacko, as the doctor thinks it would have a tendency to make him savage. The mode of capture was as follows: Ned Austin, the engineer, on coming in sight of the bluff at the eastern end of No. 4 tunnel saw what he supposed to be a man lying asleep in close proximity to the track, and as quick as thought blew the signal to apply the brakes. The brakes were instantly applied, and in a few seconds the train was brought to a standstill. At this moment the supposed man sprang up, and uttering a sharp quick bark began to climb the steep bluff. Conductor R.J. Craig and Express Messenger Costerton, followed by the baggageman and brakemen, jumped from the train and knowing they were some twenty minutes ahead of time immediately gave chase. After five minutes of perilous climbing the then supposed demented Indian was corralled on a projecting shelf of rock where he could neither ascend or descend. The query now was how to capture him alive, which was quickly decided by Mr. Craig, who crawled on his hands and knees until he was about forty feet above the creature. Taking a small piece of loose rock he let it fall and it had the desired effect of rendering poor Jacko incapable of resistance for a time at least. The bell rope was then brought up and Jacko was now lowered to terra firma. After firmly binding him and placing him in the baggage car "off brakes" was sounded and the train started for Yale. At the station a large crowd who had heard of the capture by telephone from Spuzzum Flat were assembled, each one anxious to have the first look at the monstrosity, but they were disappointed, as Jacko had been taken off at the machine shops and placed in charge of his present keeper.

The question naturally arises, how came the creature where it was first seen by Mr. Austin? From bruises about its head and body, and apparent soreness since its capture, it is supposed that Jacko ventured too near the edge of the bluff, slipped, fell and lay where found until the sound of the rushing train aroused him. Mr. Thos. White and Mr. Gouin, C.E., as well as Mr. Major, who kept a small store about a half a mile west of the tunnel during the past two years, have mentioned having seen a curious creature at

A SASQUATCH CAUGHT?

different points between Camps 13 and 17, but no attention was paid to their remarks as people came to the conclusion that they had either seen a bear or stray Indian dog. Who can unravel the mystery that now surrounds Jacko? Does he belong to a species hitherto unknown in this part of the continent, or is he really what the train men first thought he was, a crazy Indian?

We went to see Mr. McKelvie, who told us that he had checked in old records and found that all the people mentioned in the story were genuine. He had not been able to consult the New Westminster papers, published much closer to the site, because, he said, no files for that period existed. There had also been a paper at Yale itself until a few weeks before, but at that very time it was being moved from Yale to Kamloops and did not publish for several weeks.

Later I found that one man, August Castle, still lived at Yale who had been there in 1884. He said he had been only a child at the time and was not taken to see the creature but he could remember the incident well.

For nearly 20 years the story of Jacko's capture was considered to be one of the most convincing pieces of evidence for the existence of the sasquatch, but then I was told that the files of the New Westminster papers did exist after all, and what I found in them considerably weakened the story.

The *Columbian* had reprinted the article from the *Colonist* on July 5 without comment, but on July 9 the following appeared in the *Mainland Guardian*:

The "WHAT IS IT"

Is the subject of conversation in town this evening. How the story originated, and by whom, it is hard for one to conjecture. Absurdity is written on the face of it. The fact of the matter is, that no such animal was caught, and how the "Colonist" was duped in such a manner and by such a story, is strange; and stranger still, when the "Columbian" reproduced it in that paper. The "train" of circumstances connected with the discovery of "Jacko" and the disposal of the same was and still is, a mystery.

Yale, B.C. July 7, 1884.

On July 12 the *Columbian* contributed the following:

THE WILD MAN—Last Tuesday it was reported that the wild man, said to have been captured at Yale, had been sent to this city and might be seen at the gaol. A rush of citizens instantly took place, and it is reported that no fewer than 200 impatiently beg-

ged admission into the skookum house. The only wild man visible was Mr. Moresby, governor of the gaol, who completely exhausted his patience answering enquiries from the sold visitors.

It might seem that Jacko was thoroughly discredited, and yet evidence has accumulated for the defence as well. In 1970 Chilco Choate, a game guide at Clinton, B.C. wrote a letter to Dr. Grover Krantz, a physical anthropologist at Washington State university who is the only person with such qualifications who has committed himself fully to the sasquatch investigation. In the letter Mr. Choate had a reference to Jacko:

I heard this story from my father. When he first told us the story many years ago it was still quite clear in his mind and this is how it went.

My grandfather was the B. & B. engineer (buildings & bridges) for the C.P.R. when it was built west of Revelstoke. There is still a small train stop that was named after him. (Choate, B.C.) After the C.P.R. was built he became a circuit judge for the County Court of Yale, although I don't know just how long he actually was a judge. Anyway, he was there when this "Ape" was brought in and kept at Yale. "Ape" was the word my Dad says his Dad called the captive. The Ape was kept in Yale until the owner loaded it crated onto the train heading east. The story goes that he was taking it to London, Eng. to set up a side show with it and make his fortune. This was the end of the story as nobody heard of either of them again. It was either my Grandfather's opinion or Dad's opinion that the Ape must have died on the trip and was probably disposed of in any way possible. Personally I imagine it probably died at sea and would have simply been thrown overboard.

There was also a reference to memories of Jacko in a letter written to me by Mrs. Hilary Foskett, of Ucluelet, B.C.:

Did I mention that before that my mother, Adela Bastin, was educated in Yale at All Hallows in the West, a school run by Anglican nuns from All Hallows in Ditchingham, England? This school was of course a boarding school and accepted pupils from all over....

When the stories of the 'Yeti' and Sasquatch appeared in the press, Mother recalled stories of Jacko at Yale. She was probably eight or nine when she started school there and local inhabitants were still talking about the 'wild man' and the good sisters at the school took care in shepherding the pupils from school to chapel and church. In spite of this local 'fear' in her later years at the

A SASQUATCH CAUGHT ?

school Mother climbed Mt. Leakey behind Yale with a group of local people. Until well in her eighties she could recall Yale days in detail....The Dr. Harrington referred to was well known to Mother and her sisters ....

There has also been another story come to light in recent years of the capture of what would almost have to be a small sasquatch. If we had known about it back in the 1950's it might well have been possible to check on it, but it seems to be too late now. The story was told in a letter I received from a lady who had heard me talking on the radio in San Francisco. She wrote from Pacifica, California:

When I was four years old my mother died. My father placed me in a Catholic Orphanage in San Francisco, and put my brother, who was seven, with the priests of San Rafael...

When I was 12 years old, one day a nun called me and three other girls to help her. Up to the 3rd floor she took us which was the bathrooms. I believe there were 15 or 18 tubs, as this was called the 3rd Band, girls 12 and under.

We were told to draw a tub full of water. In came another nun with this creature. Sister guessed her age to be 11 or 12 years. First they drenched her hair with larkspur, as she was full of lice. Her hair was jet black, coarse and down below her hips. It took four of us to hold her in the tub while the nun scrubbed her head. She was hairy from head to toes. She just sneered at us. Once in a while she let out a gutteral sound.

After fine-combing her hair and removing lice and nits the nun took and put a large diaper on her. She tore all clothes off we tried to put on.

She had jet black eyes and was very strong.

This happened in 1916, something I'll never forget.

She stayed four days with the nuns and was taken away in a van to go East to some hospital for study, where we were not told.

We four girls were sworn to secrecy. Of the 500 children no one but us saw her.

I am now 70 years old, but can see that creature as if it were yesterday.

She stood like a human and had fingers (hairy) and toes like anyone else.

Some men who were hunting somewhere in California found her roaming the hills.

I did not give name of orphanage as I have to be sure their name will not be publicized. The nuns would not appreciate publicity. They have now changed the orphanage into a boarding school for girls.

All the nuns who were there in 1916 must be all dead by now. However they must have records of this creature being brought to the orphanage.

Sister Helena was the Superior at that time. A Sister Hortence took charge of this creature. She was in full charge of horses, cows and chickens and was a very strong person herself. She could handle anyone.

P.S. You are the first person I've told this story to.

The first thing you expect when you get a letter like that is that the name and address will be false. This lady turned out to be genuine, but when I asked for more information she wrote that she had enquired and found that records so old would have been destroyed and that no one now at the institution had any suggestion for learning anything further. The waist-length hair is not right for a sasquatch, and in general the thing does not seem very animal-like. If someone in San Francisco were to spend the time and effort they could probably get to the bottom of this story, I would think, but even if it could be proved true the creature would be long gone.

As to Jacko, whether he really existed or was just a newspaper spoof probably can't be established beyond doubt, no matter what new information may turn up in the future. In fact the story would remain very questionable even if the existence of the sasquatch as a real animal were proven. But Jacko played an important role in starting an investigation that has now outgrown any need for the story to be true.

New 'Sasquatch' found

JERRY CREW . . . something's afoot

—it's called Bigfoot

EUREKA, Calif. (AP)— Jerry Crew, a hard-eyed catskinner who bulldozes logging roads for a living, came to town this weekend with a plaster cast of a footprint.

The footprint looks human, but it is 16 inches long, seven inches wide, and the great weight of the creature that made it sank the print two inches into the dirt.

Crew says an ordinary foot will penetrate that dirt only half an inch.

"I'VE SEEN hundreds of these footprints in the past few weeks," said Crew.

He added he made the cast of a print in dirt he had bulldozed Friday in a logging operation in the forests above Weitchpeg, 50 miles north and a bit east of here in the Klamath River country of northwestern California.

Crew said he and his fellow workmen never have seen the creature, but often have had a sense of being watched as they worked in the tall timber

BIGFOOT, as the Bluff Creek people call the creature, apparently travels only at night

Crew said he seems fascinated by logging operations, particularly the earth moving that Crew does with his bulldozer in hacking out new logging trails.

"Every morning we find his footprints in the fresh earth we've moved the day before," Crew said.

CREW SAID Robert Titmus, a taxidermist from Redding, studied the tracks and said they were not made by any known animals.

"And they can't be made by a bear, as there are no claw marks."

"The foot has five stubby toes and the stride averages about 50 inches when he's walking and goes up to 10 feet when he's running."

TWO YEARS AGO reports from this area told of logging camp equipment tumbled about, including full 50-gallon drums of gasoline scattered by some unknown agency.

Bigfoot At Bluff Creek

For all that I had heard about the giant footprints at Ruby Creek I don't think it ever crossed my mind that it might be possible to see such things for myself until I read a story in the Vancouver *Province* on October 6, 1958, about a bulldozer operator named Jerry Crew, who had been seeing huge human tracks on a new road and went to the trouble of making a plaster cast and taking it to the newspaper in Eureka, California.

He certainly wasn't the first man to make such a cast—Deputy Sheriff Dunn, for instance, had made some at Ruby Creek—but he was the first to get serious attention from the press. A photo of him with the cast was carried by the wire services, and for some months the matter got a good deal of publicity.

The road where Jerry was working was being pushed towards the Oregon border in the extreme northwest corner of California. It started from the road that follows the Klamath River from the tiny town of Weitchipec to what is now Interstate 5 north of Yreka, California. The Klamath River road itself was at that time a one-lane gravel track over much of its length, and the area penetrated by the new road, up the valley of Bluff Creek, was completely roofed in with closed-canopy forest and totally uninhabited.

There were already a few access roads into the area, but they were built along the tops of the ridges where they could not be cut by slides. The new road was intended to open the country for logging, and was following the creek.

About twenty miles of road had been built when the big tracks started to appear. They were made at night, at intervals of about a week. Their basic route was across the road to get from the hillside to the creek bed, but sometimes they went for a short distance along the road and even circled the parked machines.

The tracks were all made by the same individual, and were roughly 16 inches long and eight inches wide. They were very deep in the newly-turned earth of the roadbed.

I didn't really believe it could be true, but I was interested enough to drive to California to find out.

When we got fairly close to the area mentioned in the newspaper we began stopping to ask directions, and the closer we got the more positively we were told by those we spoke to that the whole thing was a hoax and had been thoroughly explained by "the sheriff." That worthy, we were told, had discovered that it was a demented Indian boy who had been kept chained up but had escaped some years ago. He had also discovered that it was a big youth who had been in a work camp in the area and had gone wild. Moreover he had proved conclusively that the tracks were made with false feet that could be purchased in any joke shop.

Fortunately we had been exposed at home to the same easy acceptance of silly explanations for such phenomena and were aware that those explanations weren't usually very near the truth, so we kept going. At Willow Creek we met Jerry Crew and Jess Bemis, another man working on the road. Mr. Bemis guided us to the scene next morning. He left us behind on the last stretch of the twisting road and when we pulled up behind his parked pickup he had already had time to look around. He walked up to the car and said that he was sorry but the good set of tracks had just been "back-bladed" by a bulldozer that morning. I thought, "Oh, sure, just what we expected..." but he went on ..."Look around and you're bound to find some older ones."

We did look, and we did find some tracks, in the ridge of dirt by the side of the road just a few feet from the car. They were old and somewhat crumbled, but the basic shape was clear. They were very deep and they were spaced like tracks.

What particularly impressed me was the similarity between the outline of these tracks and the tracing I had of one of the Ruby Creek footprints. They did not diverge by more than half an inch at any point, and could easily have been made by the same individual.

On the way home we detoured southeast to Redding to call on Bob Titmus, a taxider-

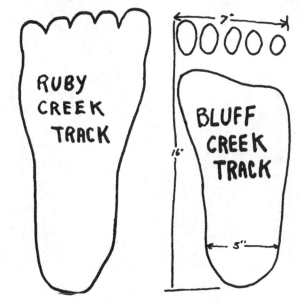

Tracings by Dunn (left) and Titmus, same scale

39

"BIGFOOT" at BLUFF CREEK

mist who was reported to have studied the tracks. He told us that over the years several hunters had told him of seeing giant human prints in various locations, but he had always insisted that they were just bear tracks. He had finally gone to see these tracks because Jerry Crew was an old friend. In fact he had given Jerry the plaster and showed him how to make casts.

One look had convinced him, he said. The tracks were genuine, although he had no idea what made them.

A couple of weeks later Bob Titmus wrote to me that he had found fresh tracks on a sandbar in the bed of Bluff Creek, and that these were made by a slightly smaller foot.

I returned to California, accompanied by Gus Milliken, from Yale, and we spent several days with Bob camped near the end of the Bluff Creek Road. Construction had stopped for the winter so there was no one else around. The tracks Bob had found were about 15 inches long and quite distinctive in shape. Punched an inch and more deep in hard-packed wet sand, they were almost perfect impressions. The sandbar was blown clean of leaves but a gravelled area surroun-

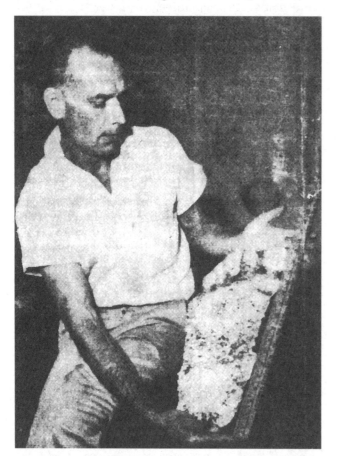

Bob Titmus of Anderson measures a plaster cast of a print reportedly made by Bigfoot. The length—16 inches.

ding it was completely covered and no more tracks could be seen. They could be found by hand, however, definite foot-shaped indentations in the gravel.

For the first time I had a chance to appreciate the tremendous pressure with which the prints are made. Where they sank an inch deep in the sand my boots made only a heel print and a slightly flattened area in the center of the soles. To make a hole an inch deep I had to jump off a log about two feet high and land on one heel.

The lower end of the sandbar was barred by a log jam and a short distance above the bar the creek ran through a little canyon with sheer rock sides. Both sides of the creek were very steep and brush-covered. There was no way any kind of machine could have been driven down to the sandbar.

The track-maker had reached the bar over the log jam and had left to climb the hillside. Most of the area was rocky or was too hard-packed to show prints. Following the line of the tracks into the bush, however, we were puzzled to find three small trees, some distance apart, each with the top partially broken off about eight feet above the ground and then twisted around the lower part of the trunk.

Most of the tracks showed a stride of four to five feet, but there was one step, down a slight incline, that covered fully six feet.

It was November when we were there, and with work on the road stopped there was little chance of anyone driving by, so I put a flash camera on a tree with a tripline crossing the road. Nothing happened while we were there but I later learned that the 16-inch tracks went down the road a day or so later, right past where my camera had been.

About that time I learned from Betty Allen, a newspaper correspondent who lived in the same town as Jerry Crew, that Indians in the area knew there were big hair-covered people in the remote valleys. But there had been no equivalent of Mr. Burns to tell the white community the story and those who were prepared to accept that the big footprints were real had at first no explanation for them.

A few people volunteered to the newspapers that they had seen a big hairy creature walking erect, generally crossing one of the mountain roads in the area at night and glimpsed only in the car lights. One construction worker who slept in a small tool shack on the Bluff Creek Road was reported to have heard something moving around outside, opened the door and confronted a manlike giant covered with hair. It was said that in his panic he picked up a chocolate bar from a table and handed it to "Bigfoot", who

Original personnel of the "Pacific Northwest Expedition" standing around a campfire near the Bluff Creek Road. From left, Ed Patrick, Tom Slick, Rene Dahinden, Kirk Johnson, Bob Titmus and Gerri Walsh, Tom's secretary. Photo by the author.

took it. I was never able to check that story because the man had quit his job and left the area.

Another man told the *Humboldt Times* that Bigfoot was only a man in a fur suit. He had seen him cross a road at night, and the suit had a droopy look around the seat. Asked whether he had seen any tracks he replied that he did not get out of his truck because "that guy was eight feet tall."

In general, however, the tracks were not taken seriously. Almost everyone was sure they were a hoax and only a few people were willing to "bite" to the extent of going to see for themselves. Among those who did have the curiosity to go and look, zoologists and other scientists were conspicuously absent.

"Bigfoot" received tremendous publicity, and as a result quite a few people spent considerable time in the area trying to catch up with him. Most ambitious project was the "Pacific Northwest Expedition" headed and financed by Tom Slick, a Texas oil millionaire. It had men in the field steadily for

almost three years. Rene Dahinden, Bob Titmus and I were all involved in the formation of it, along with the late Ivan Sanderson, a well-known naturalist, author and TV personality. Bob was field leader for the first few months, although he still had his business to run. Many sets of tracks were found as well as some hair and fecal matter that could never be positively identified, but everything that was collected, including pictures, had to be sent to Tom's Southwest Research Institute at San Antonio. Later he was killed when his private plane blew up in the air, and this material can no longer be traced. Presumably Tom's associates, who were not in sympathy with his researches of this sort, lost no time in disposing of it.

The years in which the "expedition" was active did not include any period like that in 1958 when the original 16-inch "Bigfoot" was making tracks fairly regularly in the same place, in fact most of the tracks found were of the 15-inch foot. But in 1963, when for the first time high lead logging was tried

41

Bob Titmus and Syl McCoy at Hyampom with casts of three different footprints, 17 inches, 16 inches and 15 inches in length.

in that part of the country, Bigfoot was back. Apparently the sight of logs flying up or down the hill with one end in the air really roused his curiosity, because for a while the tracks were around the machines every night.

Under the circumstances, Dave Blake, the logging operator, told me, it got difficult to keep men on the job. One evening the foreman wrote a note and stuck it on a log. When Blake asked him what he was doing he said, "I'm leaving a note for Bigfoot to come to work in the morning."

Besides the size of his feet, Bigfoot earned fame in 1958 for the feats of sheer strength he was reputed to have performed, which included throwing huge wheels from earthmovers and large-diameter culverts off the road into the creek, and tossing full drums of deisel fuel around.

I was never able to check very closely on any such stories at that time, but in 1967 I talked to four men, in three different towns, who had been working at the high-lead operation in 1963. They had all been greatly impressed by one particular incident that took place in mid morning not far from where two crews were working.

As I understand it—they did not agree in all details—there was a road crew working ahead and another crew following them installing culverts, with a trailer loaded with culverts parked by the road between the two crews. About 10 a.m. the road crew heard a crashing noise but thought the culvert crew had made it. Then at noon they went back down the road and found the trailer upside down, resting on its load of culverts. They said that it took half a dozen men to tip the load enough so that the chains holding the culverts could be released, but that Bigfoot had not rolled the trailer that way at all, he had picked it up and turned it over in the air.

"Bigfoot seemed to have a mad on for culverts," I was told by Pat Graves, who was a government road inspector on that particular job. He and the others agreed that there had been occasions when 48-inch culverts, so big they were always handled by machine, had been thrown down into the creek—the lack of any damage to the undergrowth on the intervening slope indicating that they went all the way in the air.

I have never had an opportunity to make a cast of one of the 16" Bluff Creek tracks

"BIGFOOT" at BLUFF CREEK

myself, but I have seen them several times and have copies of casts. The last time I saw what I believe was Bigfoot's track was in May, 1963, near a tiny village called Hyampom, about 60 miles south of Bluff Creek. Three sets of tracks had been found there within a few miles of each other, all apparently heading north, and Bob Titmus had made casts of all three. The biggest, at 17 inches, was the largest track we knew of at that time. All of the 17-inch tracks and those of a strange 15-inch foot with very splayed toes had been trampled by the curious before I got there, but a few of the 16 inch tracks were left and they appeared to be those of our old acquaintance.

The Bluff Creek 15-inch foot was still making tracks in that period as well, sometimes in company with a smaller one that I did not see until 1967. The tracks described in the following story from the *Humboldt Times*, August 10, 1962, were estimated to be 16 inches long, but I have seen copies of the cast that was made, and it is the 15-inch foot:

An adventurous Naval Academy midshipman, tracking Humboldt county's far northeastern wilderness area for material to complete a research paper, yesterday came back with more than he had bargained for—two plaster casts of the legendary Bigfoot's tracks.

Annapolis senior, D.M. Clark, his younger brother Danny and midshipman T.E. Williams Jr. re-kindled interest in the mysterious monster of the Six Rivers wilderness by discovery of a new set of gigantic tracks.

The two military students, both 21, and Clark's brother, 15, found the series of 16 inch tracks four miles northeast of Onion Mountain, near the Siskiyou county line. The site is about 36 miles north of the Klamath river above Weitchpec.

Clark said the tracks were along Bluff Creek, about 200 yards over a bank from a logging road on which he and the others were driving.

He said the three noticed "a disturbance in the sand" from the road and climbed down to see what it was. "At first we thought it was probably bear tracks," said Clark, "but we decided the tracks could not have been made by a bear, or anything except Bigfoot!"

Measuring with a hunting knife and the stock of a 30.06 rifle, the students guessed the length of the prints at 16 inches and the width about 8 inches at the ball of the foot.

The youths followed the tracks for about a half mile in two directions along the creek. Clark said the tracks finally went straight up a steep bank.

Clark estimated the angle of the bank at 35 degrees. Attempting the climb himself, Clark said, "no human being alive could get up it on a straight approach."

Danny Clark, a high school sophomore, went back to the car for plaster of Paris to make a print of the track.

An impression of the toes in the hillside was taken in addition to that of an entire footprint on flat ground.

Bigfoot's stride on level ground was measured at 51 and three quarter inches. Where the tracks were going up the hill, the distance between steps was 36 inches.

Clark, who said he entered Bigfoot country with a "skeptical" attitude, related last night he was 94 and 44/100th's per cent convinced of Bigfoot's existence.

"And the remaining per cent is spent wondering how anybody could fake the tracks," he said. "Reading about this, it's hard to imagine it because of all the other things it could possibly be.

After seeing for himself, the naval student changed his mind. "Nobody could pay me to walk around in that country with fake 16 inch feet, especially carrying 600 pounds of rocks to make deep impressions," he stated.

Referring to the depth of the tracks, Clark said they were sunk into the ground about an inch and a half deep. Clark, who weighs 185 pounds, jumped from a five-foot rock and sunk his heels only a quarter of an inch into the ground.

The tracks were found at about 5:50 p.m., only two hours after the party of three arrived in the area on their first trip.

A steady rain began to fall at about 7 o'clock and continued through the night. Unable to move their wet plaster casts, the resourceful adventurers hollowed out the top of a log and covered the casts with wood and brush. They then built fires around the log so the plaster would dry. Not wanting to leave the area, they went about two miles back down the road and slept in an empty surveyors' camp.

When the plaster impressions were dry enough, the party drove back to Willow Creek.

Without footprints the sasquatch would present no problem. They could be dismissed as a mixture of legends and lies, with perhaps an occasional hallucination thrown in. There would still be a movie to explain away, but that would be an isolated problem.

Footprints, unfortunately, have no such easy explanations. They are real, and they have to be made somehow.

There are only two possible explanations for them—at least I have never met anyone who could suggest a third. Either they are

"BIGFOOT" at BLUFF CREEK

made by men or they are made by some creature with feet of the right size and shape. Such a creature would have to be very large, yet unknown to science, and a close relative to man. The very idea is ridiculous. Alternatively, there must be a widespread, well financed, continuing organization devoted to the faking of big tracks all over western North America and a lot of other places. That is also ridiculous. The only reasonable explanation is that it is all some sort of mistake and the tracks don't really exist.

That approach to the problem has just one serious weakness. The tracks DO exist. something goes around making enormous, humanlike tracks. That is a fact which the scientific world has so far chosen to ignore, but which it is in no way able to explain.

The footprints may be close to 20 inches in length and range from there down to a size where they can no longer be definitely distinguished from human footprints. They resemble the print of a human foot in that they are "plantigrade," that is the entire foot from heel to toe rests on the ground, and that they have a big toe on the inside which is considerably larger than the other toes and extends beyond them.

They differ from the normal human track in that they are flat, in most cases without any sign of an arch, and that their width

"It's just some guy with a size 14 shoe."
Well, here's a size 14 shoe beside a cast of a 17-inch footprint.

is very great in proportion to their length. A typical print 16 inches long is almost eight inches across the ball and five inches across the heel. A shoe manufacturer has been quoted as saying that one particular 15-inch print would require a size 21 shoe for length but the width would have to be 13 sizes greater than the widest shoe made. A foot of normal human proportions 16 inches long would probably not be more than six inches wide, although the print of a person customarily going barefoot might tend to be wider and flatter than that of a person wearing shoes.

The shape of the big footprints is not particularly consistent, however. Most of those I have seen are sufficiently distinctive in shape so that they would be readily recognizable even if all were the same length. Some, such as a 14½-inch track cast in northern California in October, 1967, are proportionately narrow. One 15-inch track has the toes widely fanned out. Two fairly short tracks with narrow heels are as wide as the largest tracks at the ball, and have large toes. There was one track in the Bluff Creek area that had a very deep arch. For size and shape it could have been a human track, since it was only 11 inches long, but it appeared alongside much larger tracks, and the depth of impression indicated that whatever made it was very heavy.

Some tracks are reported to show three, four or six toes. I have seen three and four-toed casts, but all the tracks I have seen in the ground had five toes, although one print in a trail of five-toed tracks showed only four of the toes.

An unusual feature discernable in most of the Bluff Creek tracks, in varying degrees, was a division right in the middle of what appeared to be the ball of the foot. Ivan Sanderson concluded from examination of one set of casts that the foot had very long but partially webbed toes, with a pad under the first bone of the big toe. His theory may well be correct, since human toes and fingers both have considerable webbing in front of their base joints, and on the human hand calluses form primarily at the top of the web rather than over the joint. Alternatively, the callus over the joint on Bigfoot's big foot may simply be so thick it has to split in order for the foot to bend. So heavy a creature travelling barefoot over rough terrain would have to have extremely heavy pads on its feet. Some picture-perfect prints inspected by Bob Titmus near Terrace, B.C. in 1976 had distinct pads that not only differed from any of the prints at Bluff Creek, they differed markedly on the left and right feet.

"BIGFOOT" at BLUFF CREEK

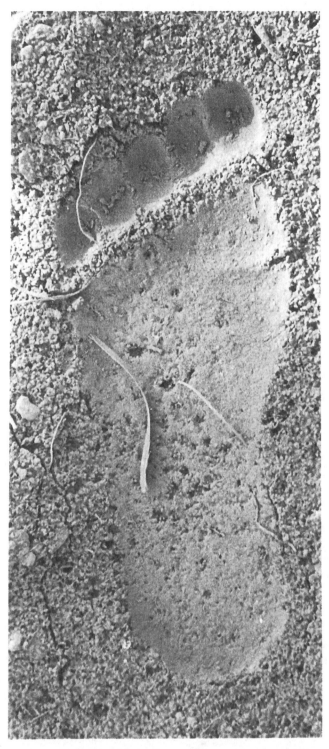

The best photo of a sasquatch footprint that I know of, this shot by Rene Dahinden shows a 13-inch print in deep dust dampened on the surface by a brief rain. Note the rounded toe prints, the deeper impression at the heel and the split in the ball of the foot. Behind the toes and in the cleavage in the ball the raindrop pattern is undisturbed. There are pressure cracks around the heel and in front of the toes.

The most impressive thing about the big footprints is neither their size nor their shape, but the depth to which they are impressed in the ground. This must vary, of course, in relation to the type of material in which they are made. In loose dust or snow over hard ground they may be no deeper than the print of a human foot, but in mud, dirt, sand, hard snow and sometimes even in gravel, they will sink to depths far greater than a man's footprint. I have seen 15-inch tracks an inch deep in hard-packed wet sand where my own shoes did not leave a complete track at all. False feet worn by a person would show the opposite effect, tending to sink in less than the normal amount, as with snowshoes. Many of the big tracks have more than three times the area of the normal human foot, as well as sinking in to several times the depth.

As far as I know there has never been a systematic study of a set of these big prints by anyone specializing in the relationship of weight to compression of the ground, although this is a field in which a great deal of specialized knowledge should be available, if not among students of tracks then among civil engineers. I was present when a university physics technician examined a few rather poor tracks and concluded that they could not have been made by a weight of less than 300 pounds. To my own mind, I could not accept that some of the tracks I have seen could be impressed by a weight of less than half a ton.

On this point, the following story appeared in the *Humboldt Times* on November 18, 1958:

Another "Bigfoot" discovery broke in the headlines in San Jose yesterday from Los Gatos, where Dr. R. Maurice Tripp, geologist and geophysicist, completed the cast of a 17-inch footprint he found in the Bluff Creek area of Humboldt's storied wilderness.

Dr. Tripp reported that his engineering studies of the soil properties and depth of the footprint from which he made the cast indicate the weight on that huge puppy to have been more than 800 pounds.

The print is distinctly different from that of a bear or any other animal known to be in the area," Dr. Tripp told the San Jose News.

The story admitted that although Dr. Tripp did not discount the chance of fraud, he believes that the tracks lend credibility.

"Several people have tried to track him and in one instance his footsteps could be followed a distance of a mile and a half through brush country," he said. "It would be difficult to fraudulently prepare hundreds of such tracks overnight—particularly in the

45

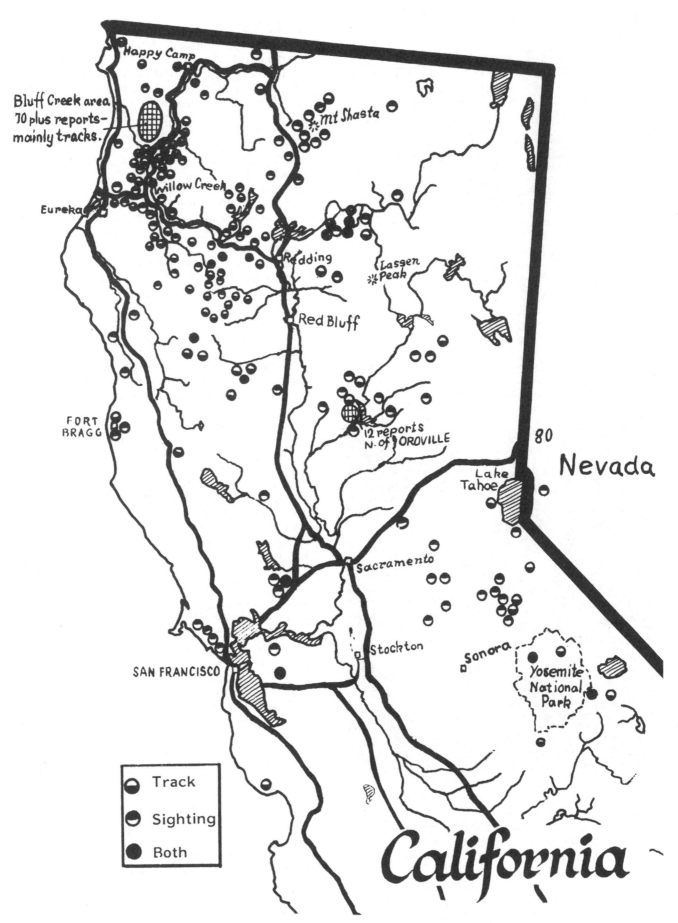

Bluff Creek area
70 plus reports—
mainly tracks.

Happy Camp

Mt Shasta

Willow Creek

Eureka

Redding

Lassen Peak

Red Bluff

FORT BRAGG

12 reports
N. of OROVILLE

80

Nevada

Lake Tahoe

Sacramento

Stockton

Sonora

SAN FRANCISCO

Yosemite National Park

California

	Track
	Sighting
	Both

SIGHTING AND TRACK REPORTS IN NORTHERN CALIFORNIA

"BIGFOOT" at BLUFF CREEK

type of country in which they were found."

He mentioned other evidence of Bigfoot. In the area was a strand of hair found on a tree seven feet, three inches above the ground. Dr. Tripp said the hair was discovered immediately after reports had been received that the mysterious monster was in the area.

Dr. Tripp said his interest is both scientific and the result of curiosity. He first became interested 18 months ago when he received a report that a clergyman and a woman of Bluff Creek said they actually had seen the elusive "Bigfoot." He was able to reach the region in time to get a cast of the footprint.

"Now we just have to find the foot that fits it," he observed.

Numerous expeditions have ventured into the Bigfoot country, at least two of them conducted on a scientific basis. Many others were individual efforts, but none has come up with any specific proof either in photographs or by contact with the mammoth mystery man of Humboldt. Indians in the area say the legend goes back to about 1850. Dr. Tripp is the first investigator of scientific note to estimate "Bigfoot's" avoirdupois.

In a sense the footprints are physical evidence. They can be photographed, measured and tested for compression. Plaster casts of them can be taken that must closely resemble the feet that made the prints. On the other hand, footprints are not anything more than a shape bounded by the surface of some material, and they could be made by removal of the material or by velocity of impact as well as by weight. Persons who have studied the big prints in the ground have always, in my experience, concluded that they are made by weight, but I do not know that this has ever been established by scientific tests, which I am sure could readily be done.

Prints dug out or imprinted by velocity would almost certainly be of human manufacture, but if imprinted by weight they could be made by humans only with mechanical assistance.

With that in mind the following story, printed October 15, 1958 in the *Humboldt Times,* is of interest:

The story of ancient and mysterious giant sized tracks is an old, old story, Dr. T.D. McCown, professor of physical anthropology at the University of California said yesterday.

Speaking of Bigfoot's gigantic 16-inch tracks which have appeared on the Bluff Creek access road construction job, Dr. Mc-Cown said that such tracks have been repor-

ted since the beginning of time.

He said that records show reports of footprints, most of them the same size as the ones found in Humboldt County. Such records indicate that millions of huge tracks have been found through the years. He did not specify their origin.

Most of the track reports have come from Africa and Asia, although many have been reported in North and South America. Some tracks have also been found in Europe, the professor said.

If the Humboldt County Bigfoot is tracked down and discovered it will be the first time in history that the mystery has been solved, Dr. McCown said.

He went on to say that there have been also many reports of tiny footprints supposedly made by little people.

I have seen tracks myself that were far too deep to have been made by humans wearing false feet. I have seen them at the bottom of a steep gully where no heavy machinery could have been brought in except by helicopter, and I have seen 600 yards of tracks made in the short period when the top layer of deep dust was damp after a sprinkle of rain. None of the tracks I have seen appeared to have been made by digging them out, and in fact the tracks in dust with a skin of mud could not have been made that way. Nor could they have been made by any kind of rigid mold, since they showed numerous distinct variations, particularly in toe positions, from print to print. They could not have been made by anything mounted on a wheeled or tracked vehicle, as they were able to climb banks with the feet still horizontal.

I have never seen a set of tracks deep in the bush, or followed the trail by other signs on ground where clear tracks could not be seen and found the tracks again on the line I was following, but I have talked to other people who have done these things.

Once the idea of bears is disposed of no one has been able to propose any explanation for these tracks beyond the two obvious ones, that they are manufactured by humans as a hoax or that they are made by an enormous biped with humanlike feet. Superficially the hoax is by far the more acceptable idea, but it has difficulties.

For the years before the general public was aware of "Bigfoot" there is no apparent explanation as to who would be doing the hoaxing, or why. Considering the time and the distances involved it would need to be the work of an organization rather than an individual. Still, people do strange things for no apparent reason, so perhaps the who and why should be ignored. The real diffi-

47

"BIGFOOT" at BLUFF CREEK

culty is how. In decades of wrestling with the problem no one to my knowledge has ever devised a way to manufacture a complex set of tracks. It is possible to duplicate a cast by pressing it in the ground, and one successful hoax was carried out with a refinement of that technique, but only with a small number of tracks on level ground.

In the early days at Bluff Creek a popular TV program offered $1,000 to anyone who could tell how the "Bigfoot" tracks were made, and there were a great many suggestions submitted, but none proved effective. The weight would obviously require a machine, and the machine would have to travel on feet, without leaving any other kind of track. It would have to be able to climb and descend steep banks, vary the length of stride and the placing of its weight on the feet, and move the toes freely.

Another problem would be doing the faking without getting caught. When thousands of sets of tracks have been made, many of them in locations where someone might happen along at any time, it is inevitable that the hoaxers would many times have been caught or at least interrupted at their work. Although there certainly have been some hoaxes there have been no instances that I am aware of where people were caught making tracks, and only a few reports of tracks stopping where they should have continued—something I have never seen myself. On the other hand dozens of people report seeing a big hairy biped making the tracks or seeing the animal and then finding the tracks where it had been.

The idea of a complex hoaxing organization making big tracks all over North America—one scoffer even suggested that it included "hoaxers hoaxing other hoaxers"—surely does not recommend itself to any reasonable person, no matter how certain they may be that no big undiscovered animal exists. The only logical conclusion they can reach is that the tracks cannot possibly occur, but they do.

One particular track that has received a great deal of attention is that of "old cripple foot" in the area north of Colville, Washington. Found late in 1969, in both dirt and snow, those tracks showed a fairly normal five-toed left foot, 17 inches long, but the right foot was twisted, with only four toes showing and with two large nobs on the outside edge.

When zoologists are asked about the tracks they can be counted on to suggest that they are made by bears, and I have often listened to explanations of how a bear's hind feet may overlap his front feet, leaving what appears to be a bipedal track. It is true that a bear's foot is plantigrade, but otherwise his track has little resemblance to the big prints. His longest toe is in the middle, and his outside toe, the "little toe" of the human foot, is larger than the inside or "big" toe. He cannot make a deep track without leaving claw marks, and there are no bears with 16-inch feet this side of Alaska, if there. No print made by overlapping feet would stand close examination, since a bear's front foot is much smaller than his hind foot, but carries the greater part of his weight and therefor sinks a good deal deeper. An overlapping track would have to be very indistinct not

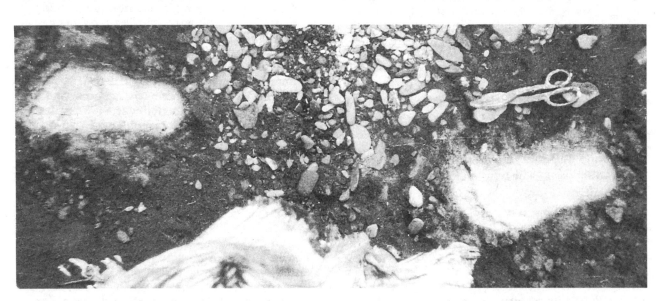

Two of the Bluff Creek 15-inch tracks on a gravel bar, sprinkled white for contrast. This is the type of track first found by Bob Titmus and the type cast by Midshipman Clark. Stride here would be about 40 inches. Photo courtesy Al Hodgson.

"BIGFOOT at BLUFF CREEK

The oldest track photo in my files, taken in 1947 on a utility right of way between Eureka and Cottonwood in California. Photo courtesy of Bob Titmus.

to be recognized as such, while many of the "Bigfoot" tracks studied have been beautifully clear in every detail. Of course once a zoologist has seen a cast or a photo of a track the "bear" explanation is heard no more.

The only other animal with a large plantigrade foot would be one of the known apes, but all of these have an entirely different big toe, set off from the other toes like a human thumb. Also, big apes when walking are basically quadruped, using the knuckles of their hands as well as their feet. A final possibility would be a bear walking on its hind legs, but while a bear in the wild will rear up on his hind legs he will not walk that way to go anywhere, and if he did he could not take long strides.

This brings up the final feature of the big tracks. They show a normal walking stride of three to four feet, not out of the question for a tall human, but they can reach six, seven or even ten feet where they appear to be hurrying. The people who saw the tracks at Ruby Creek all agreed that the sasquatch took a four-foot railroad fence in stride. The same tracks crossed a potato patch and left the potatoes crushed under the ground—a feat that I thought unbelievable at the time. I have since been told that big bears can do the same.

There is simply no way of guessing how often the big tracks are seen, but they are more common than is generally realized. Big and obvious as they are, they are often missed even by people who know where they are and are looking for them. Few people who do see them know where to get in touch

with anyone who is investigating these matters. None of us were notified in 1963 when Bigfoot was making tracks night after night around the logging equipment above Bluff Creek.

As of May, 1980, I had on file almost 200 track reports from northern California alone, going back as far as 1870. The oldest photograph is from 1947. Having seen some of the things enthusiastic people will call tracks nowadays I am well aware that there must be many modern reports in the file that shouldn't be there, but I have a lot of faith in the old ones from the days when a person would tend to find other explanations for any track that was not very definitely something unknown.

Two physical anthropologists have been greatly impressed by the anatomical accuracy of the crippled track, so that in a substantial way it is the best-authenticated of all tracks. Unfortunately it is also associated with the most elaborate hoax of all, a yearlong buildup which eventually led to the production of a more than questionable film. It is certainly possible that they were genuine tracks even though they started the series of events leading to the film, but it is also possible that they were faked by someone with a knowledge of foot structure.

In general it is my opinion that tracks are very strong evidence that an unknown animal does exist, and even if that is not accepted they are in themselves a phenomenon that would be well worth the expenditure of time and money for a full scientific investigation. They are not imaginary. Something of interest to mankind is definitely going on.

Apes In The Islands

In the early days of the California investigation we had accumulated far more reports of tracks than of sightings of the sasquatch themselves, but that situation soon changed, starting with a letter from Ivan Sanderson asking me to check on a report he had received that people in an Indian village on the B.C. coast called Klemtu encountered apelike creatures quite frequently.

Klemtu is on an island only about 15 miles long, so it seemed that if anything the size of a sasquatch was wandering around there it could easily be found. When Bob Titmus and I arrived there, however, we found that the island, and all the country around, was a mass of mountains covered with fantastically thick vegetation. Things like salal, which I was accustomed to seeing a foot or so high, were solidly matted to heights of seven feet and more. We learned that the Indians never travelled inland—even in the old days there had been only one trail on the island, across one end of it. They went everywhere by boat, even doing their hunting and trapping mainly on the beaches.

We learned that Indians all along the coast, not just at Klemtu, were quite familiar with what they referred to as "apes." Joe Hopkins told of seeing one on a clam beach south of Klemtu just a few months before. The year before that his brother encountered one on another island while deer hunting. Two other Klemtu men had recently shot at one on shore from their fishboat. At Bella Bella and Bella Coola also we found that a few minutes of making enquiries would turn up someone with an ape story to tell.

Footprints were not so common, as there was little ground suitable to show them, but we were told that a crew of men from Klemtu who were building a dam on a lake above the village had seen about a dozen sets of tracks crossing the beach at the head of the lake, and while one set was very large there were some little ones as well. That was the first report we had of what might be a family group.

The apes were reputed to eat shellfish in the winter, and they were accused of throwing rocks from concealment on some occasions when groups of people were clam digging on remote beaches at night. The rocks were not thrown straight, but dropped down howitzer fashion near the lanterns. When that happened the Indians would leave.

The situation was so interesting that Bob spent several years scouting the area by boat, with financial backing from Tom Slick until his death and later on his own. He found footprints on several occasions, but lost his collection of casts when his boat burned and sank. Once the prints were on a tiny island a considerable distance offshore, confirming the Indians' claim that the apes swam from island to island. Another time they were in snow on the boardwalk behind the Klemtu cannery.

Sightings have been rare on the islands and the adjoining coast in recent years. On a trip along the coastline all the way to the Alaska boundary in the summer of 1975 I did not learn of any reports from the past five years and I have heard of only one since. There was still lots of activity while Bob was there, however, including two separate incidents in which groups of people in boats watched apes on the shore. Near Butedale, north of Klemtu, a man out fishing in a small boat told Bob that he saw two of the creatures on a rocky islet and a third one swimming towards the islet from the mainland shore. Some of the witnesses were Indians, as are most of the full-time residents of the coast, but others were non-Indian residents or visitors.

One of the people most interested in the ape reports was Clayton Mack, at Bella Coola, a famous game guide specializing in grizzly bears. He told us of something that had happened to him almost 20 years before. I did nothing about it then, but in 1967 I made a tape recording of his story. Here is the text of our interview:

Q-What is your occupation?
A-Guide, fisherman.
Q-Guide for what?
A-Grizzly bears.
Q-And how long have you been a grizzly guide.
A-About twenty-five years.
Q-Have you any idea how many bears you've shot in that time?
A-No.
Q-How many do you get in a year?
A-Oh, the most I've got is about eighteen.
Q-So you'r talking in the hundreds, anyway.
A-Oh yes.
Q-So you know quite a bit about grizzly bears.
A-Well, I know a little about them.
Q-Can you remember what year the incident you were telling me about took place?
A-I don't quite remember, but it was some time in the Forties.
Q-And where did this take place?
A-Jacobsen Bay, about twenty miles west of Bella Coola.
Q-Where were you, and what were you doing?

APES IN THE ISLANDS

A–I was going into Quatna to do some dog-fishing there.

Q–You were on a boat, were you?

A–On a boat, yes, when I saw this thing.

Q–And how far would you be from shore?

A–I was about four hundred yards from the beach when I first saw it. Now I thought it was a bear the whole time, so I nosed the boat right towards him. I have this old boat you see, it goes putt putt putt, makes a lot of noise. So I was heading towards him—I didn't have any binoculars but I had a rifle—and I saw him stand on his hind feet. He didn't go down on his four legs, just stood on his hind feet, till I got about 300 to 250 yards away. Then he raised, kind of looking at me. Then he straightened out, like he was doing something, and he started walking up the sand. He saw the boat coming towards him and he turned; he didn't run, he just walked. But halfway up to the timber from where I first saw him, he stopped and turned and looked back at me, without turning his whole body towards me. That sure puzzled me at the time, because bears don't do that, bears don't look back like a human being and look back at you. They turn their whole body towards you, and they go down on their forelegs, but this thing didn't do that, he just kept walking on his hind legs till he hit the timber.

Q–And what did it do after that?

A–It just turned towards the timber again and kept walking. It walked on top of the drift logs on the beach. It stepped on top of them without going on its forelegs. It was doing that until it hit the timber. Then I could see the short young growth timber that's about ten feet high. He went right through that thick part, and I could see the tops of those little trees, it looked like he grabbed hold of them as he went through it, because it moved quite a bit, like he was spreading them.

Q–Spreading them like a man would with his arms?

A–Yes, like a man. That's all I could see of him.

Q–During all the years you've hunted grizzlies, have you ever seen one walk that far on its hind legs?

A–Never in my life.

Q–But this one, on the other hand, walked for quite a few yards uphill—

A–Sixty-five to seventy yards.

Q–...and then stepped up on some logs, still on his hind legs?

A–Yes, on his hind legs.

Q–And at the time you didn't know anything that this could be but a bear?

A–No. I heard about these sasquatches ever since I was a kid, but I just didn't believe in them, you know, but now I've been thinking I believe it. At the time I wasn't too sure, but I told other people, old people, and they said that was a sasquatch.

Q–But at that time you didn't know anyone else who'd actually seen one?

A–Well ....saw the same thing the same year.

Q–But you didn't know about it then.

A–No, he didn't tell anybody about it, because people would say, "What kind of a drink have you been drinking?" That's the first thing they say when you tell stories about these things.

The name omitted is that of one of several people in Bella Coola who spoke up about having seen sasquatches only after Clayton had broken the ice. Like many other people who have given me information he does not wish to be identified in print. Another man at Bella Coola with an interesting story was Jimmy Nelson, a young logger and fisherman. This is what he had to say:

Q–Can you tell me when this incident took place?

A–In November sometime, when I was out hunting.

Q–And what year?

A–1965.

Q–Where were you hunting?

A–At Green Bay, just a few miles from Bella Coola.

Q–Were you up the hill?

A–Yes, I was about half way up the mountainside.

Q–And how high would the mountains be around here?

A–I don't know exactly how high they'd be.

Q–Well, they'd be about five to seven thousand feet anyway, wouldn't they?

A–About that, I guess so.

Q–So you'd be quite a ways up.

A–Yes.

Q–What exactly were you doing at that time?

A–I was starting back down, in the middle of the slashing.

Q–Had you been through that logged-off, slashed area before?

A–Yes.

Q–Now, what did you see?

A–Sort of a black form; it looked like a man....with a white streak on the front below his head.

Q–What was he doing?

A–He was walking up the slashing.

Q–Did you get any idea of how big he was?

A–Well, he seemed quite big, because I was crawling under the logs and trees that were cut down by the loggers—I had to climb over

51

and crawl under some, but he was just walking over them.

Q-How long did it take him to go across the slashing?

A-It took him about 10 minutes.

Q-And how long did it take you?

A-About two hours.

Q-About how far away would you say he was?

A-Two hundred yards.

Q-Could you see his face at all?

A-No.

Q-You said he was black colored; could you see whether this was his skin, or whether he had hair or clothes or anything?

A-No.

Q-It walked like a human, did it?

A-Yes.

Q-About how long were you watching?

A-About ten minutes.

Those who are familiar with the "slash" left behind by loggers in the coast forests will realize the tremendous physical ability involved in crossing in 10 minutes an area it took a young logger two hours to cross. That is not the whole of it, either. Jimmy Nelson told me that he crossed the slashing at right angles, going straight up, but the thing he watched ambled across it at a considerable angle, starting well to his left and crossing over until it entered the trees at the top of the logged-off area far to his right.

The most recent British Columbia report on file as this book is written is from the Bella Coola area. Four young men fishing about four miles from the mouth of the Bella Coola River on February 23, 1980, found tracks in a patch of snow on the ice, 21 inches long and 12 inches wide. Clayton Mack was called and he took pictures of the tracks, but I have not seen them yet.

The best tracks ever cast in British Columbia were found near Terrace, by a slough adjoining the Skeena River, in July, 1976.

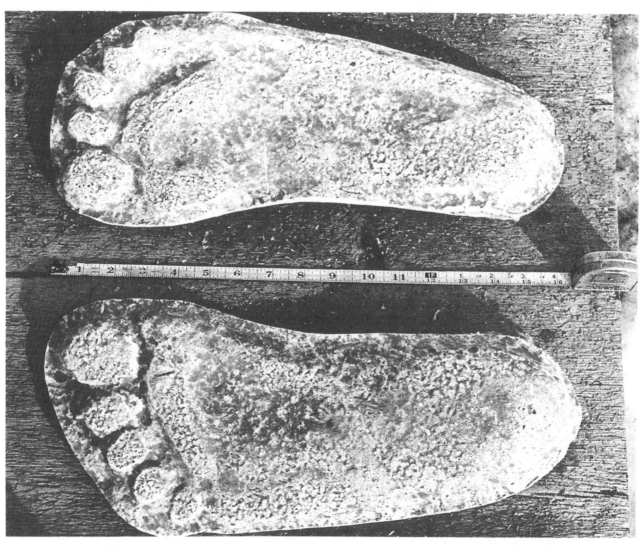

The Terrace casts. Note the great difference in the pads behind the toes.

APES IN THE ISLANDS

They were found by some children, and by good luck a man they told about them, Dick Bates, was a friend of Bob Titmus. Bob, who was then living at New Hazelton, got to Terrace the day after the tracks were found and made casts of five of them. Here is part of what he had to say about them:

The tracks measured 16½" x 6½" and the stride appeared to be a normal, unhurried walk and measured on an average 78"! This track (referring to one particular cast) was in a hard, moist clay and silt mixture soil. The tracks were impressed up to 1¼" deep, whereas my own boots left scarcely any mark whatever. It is my opinion that this individual creature could weigh well in excess of 800 pounds and probably stands over 8' tall. The tracks were approximately four days old and had been rained on 2 or 3 days before I saw them. This washed silt into them leaving flat spots in the toes and some of the calloused regions which show badly on the casts. I now deeply regret not removing the silt before casting them. (But no one likes to do anything that could be considered tampering with a track before casting it.) ....Many branches, up to the size of my wrist, were broken, or twisted off, generally 6' above ground, where this creature moved through the bush....

The bracketed remarks are the author's. In spite of the flattening caused by the silt the casts are exceptionally good ones, comparable with the best from Bluff Creek.

Presence of the "apes" on the islands along the Inside Passage would indicate that they must be good swimmers—a most unlikely attribute for an ape—but it would take more than that to explain how they could get to the Queen Charlotte Islands, almost 50 miles from the mainland, and I don't think any of us ever considered such a possibility. On June 24, 1970, however, the Vancouver Sun had the following story:

QUEEN CHARLOTTE ISLANDS—These isolated islands 400 miles northwest of Vancouver have one thing in common with other areas on the west coast.

They have a sasquatch.

Not to be outdone by Harrison Hot Springs, Alert Bay, California and Washington, the Charlottes have found a hairy creature to call their own.

Several sightings on Graham Island—the northern island of the chain—have been reported in the last year.

Tina Brown, 21, Of Skidegate Mission, said she saw something which looked like a sasquatch about four months ago at a church camp seven miles north of Skidegate.

"It was standing there in the bush and our headlights went right on it, then it walked away into the bush," she said.

Miss Brown said the creature was about seven feet tall, had hair all over and small beady eyes.

She said it had hair on its face and looked more like a gorilla than a man.

Miss Brown said she and a friend, Herman Collison, were only about 20 feet away

They said it was about midnight and "I was really frightened."

They called the police, who went to the area but found no trace of the creature.

A second sighting was reported a short time before by a boy near the logging camp of Juskatla.

That time the sasquatch lifted up its arms and ran. Some footprints were found, Miss Brown said.

Another sighting was reported more than a year ago near Queen Charlotte City.

Miss Brown said there is an Indian legend that when a person sees a sasquatch, he or someone close to him will die within a year. Miss Brown said the man who spotted the one near Charlotte City was killed in an accident exactly a year later.

In January of the following year Bob Titmus was notified of another sighting on the Queen Charlottes, this time by four deer hunters, only one of whom was an Indian. Snow had fallen in the meantime, but he was able to find three partial tracks and one good one, indicating that the sasquatch have indeed made the crossing to the island. How any large animals got there is a mystery. Some did and some did not. The wolf and the cougar are among those that never got to the islands, but the black-tail deer and the black bear did.

Indian accounts of sightings along the coast go back a long way. A man named Billy Hall, from Kemano, is said to have shot a sasquatch in 1905. No contemporary record of that incident has been found, but there is one reported in the Vancouver *Daily Province* two years later, March 8, 1907. This is the story:

A monkey-like wild man who appears on the beach at night, who howls in an unearthly fashion between intervals of exertion at clam digging, has been the cause of depopulating an Indian village, according to reports by officers of the steamer Capilano, which reached port last night from the north.

The Capilano on her trip north put in to Bishop's Cove where there is a small Indian settlement. As soon as the steamer appeared in sight the inhabitants put off from the shore in canoes and clambered on board the

APES IN THE ISLANDS

Capilano in a state of terror over what they called a monkey covered with long hair and standing about five feet high which came out on the beach at night to dig clams and howl.

The Indians say that they had tried to shoot it but failed which added to their superstitious fears. The officers of the vessel heard some animals howling along the shore that night but are not prepared to swear that it was the voice of the midnight visitor who so frightened the Indians.

Bishop's Cove is not to be found on modern maps or in sailing directions, but I was recently able to establish that it is on the mainland coast about 30 miles north of Butedale. No one lives there now.

At about the same period that we were starting the search on the coast, an outstanding sighting report came from another area altogether, the southeast corner of British Columbia. Bob Titmus and I interviewed the witness some time later, but his story is well told in the original report published in the Nelson *Daily News*, October 4, 1960:

Man or beast, or both? Whatever it was that sent John Bringsli of Nelson fleeing in blind panic from the head of Lemmon Creek, hurling his huckleberry pail into the bush and racing for home in his early-model car, it had pulled a speedy disappearing act by the time he and a group of hunters returned to the scene.

Mr. Bringsli, woodsman, hunter and fisherman in the Kootenay district for more than 35 years, swore on his reputation as an outdoorsman that it was "definitely not a bear."

In an interview, Mr. Bringsli related his experience with an "unknown creature" seen while on a huckleberry picking expedition alone near Six-Mile, and unashamedly told of his frantic race over 100 yards of stunted bush and dwarfed underbrush to his car.

I had just stopped my 1931 coupe on a deserted logging road a couple of weekends ago and walked about 100 yards into the bush. I was picking huckleberries.

"I had just started to pick berries and was moving slowly through the bush. I had only been there about 15 minutes.

"For no particular reason, I glanced up and that's when I saw this great beast. It was standing about 50 feet away on a slight rise in the ground, staring at me.

"The sight of this animal paralysed me. It was seven to nine feet tall with long legs and short, powerful arms, with hair covering its body. The first thing I thought was, 'what a strange looking bear.'

"It had very wide shoulders, and a flat face with ears flat against the side of its head. It looked more like a big hairy ape.

"It just stood there staring at me. Arms of the animal were bent slightly and most astounding was that it had hands, not claws.

"It was about 8 a.m. and I could see it clearly," Mr. Bringsli said.

"The most peculiar thing about it was the strange bluish-grey tinge of color of its long hair. It had no neck. Its apelike head appeared to be fastened directly to its wide shoulders."

Mr. Bringsli stood with mouth agape staring at the thing for about two minutes. Then it began to slowly walk, or rather shuffle, towards the paralysed huckleberry hunter. It was then that Mr. Bringsli decided it was time for him to find another berry-picking location.

He sprinted the 100 yards to the car and drove recklessly down the old logging road and home.

Mr. Bringsli returned to the scene next day with a group of friends armed with high-powered rifles and cameras but the strange beast did not reappear. They did find one track nearby. It was from 16 to 17 inches long. There were no claw marks but rather a "sharp toe" print as described by Mr. Bringsli.

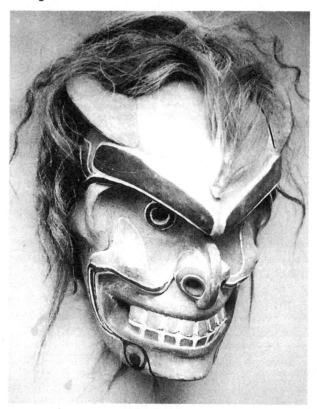

This Kwakiutl Indian dance mask in the collection of the British Columbia museum represents the Wild Man of the Woods.

The Quiet Years

The years when tracks were first appearing frequently in northern California and on the north coast beaches of British Columbia saw considerably more sasquatch activity than those that followed. Tom Slick, who was the only person putting substantial amounts of money into the search, died in 1963. Rene was relatively inactive for several years after a disillusioning experience in California. Bob Titmus and I also had to spend most of our time earning a living. We were all so out of touch that we did not even learn of the new epidemic of Bigfoot tracks at Bluff Creek in 1963 and 1964. There were not nearly as many other people doing any investigating in those days as there are now, but there were some. My own activity had been confined mainly to British Columbia and California, but things were also happening in between.

In October, 1959, Bob had investigated a remarkable incident in southern Oregon. First word of it was a newspaper account, the following being the version printed in the Portland *Oregonian* on October 23:

ROSEBURG (Special) — It couldn't have been an abominable snowman, because it was raining at the time, but two boys told police here Wednesday that they saw a 14-foot man-like creature stalking through the woods near Tenmile, about 15 miles southwest of here.

In fact, one of the boys took five shots at "the thing," as officers labeled it. Police didn't name the boys, aged 17 and 12.

The youngsters said they saw "the thing" twice, once last Friday and once Monday. The boys, who said they saw it from about 50 yards away, described the creature as being covered with hair, walking upright and having human characteristics.

State Police Sgt. Robert Keefe, Roseburg patrol supervisor, said the boys related they didn't tell their parents about it last Friday "because we didn't think anyone would believe us."

They went back Monday to the clearing near an abandoned sawmill where they first saw "the thing." Sure enough, it was there again. The older boy foresightedly had taken along a .30-caliber rifle and fired five shots from less than 50 yards, he told the officers.

"It ran off screaming like a cat, but louder," the youth said.

The youngsters said they then found humanlike tracks 14 inches long. Police looked too. The footprints are large, they agreed.

Sergeant Keefe said he had one of his game officers check the tracks. "He said it looked

like a bear track or something that resembled it," said the sergeant. "There isn't any doubt in our mind that it was an old black bear."

But the sergeant's skepticism didn't speak for the hunters of Roseburg, and Thursday afternoon two parties of them were out working the area with their dogs and rifles.

"Besides," added the sergeant in his skepticism, "they said they shot the thing with a 30.06, 180 grain soft-pointed ammunition. I don't know anything that wouldn't stop —unless it would be an elephant," he said.

"It's like I heard one of the guys out there say, "Well, gee, I think it's time we telephone the papers and tell them the flying saucers are around again," Keefe said with a laugh.

But when he asked the boys: "Could it have been a bear?" the boys replied that it couldn't have been; that they had seen bears before.

Besides, those—brr-r-r-r—footprints showed five toes and no claws. Police said they would continue the investigation.

Bob Titmus was on the scene a few days after the incident, and he found that the newspaper had missed some important points of the story. Only the younger boy had seen the creature on the first day, and the 17-year-old he had recruited to accompany him in looking for it again was a hunter, in and out of season, who reportedly always shot deer in the head so he wouldn't spoil any meat.

On the second day the boys saw the animal down below them in a valley and it saw them up on the ridge. It immediately came after them, appearing on their level with startling speed, but then approached them slowly, swinging outstretched arms as if it sought to herd them ahead of it along the ridge.

They ran, but the older boy paused several times to pump 30.06 slugs into its chest. He could see the impact as they hit the animal, and once it went down until its knuckles hit the ground, but it kept coming and did not seem to lose its composure. They were running and did not see what it did when it stopped following.

Bob checked the area and found tracks that confirmed the boys' story of their own and the animal's movements, but the tracks of the creature did not resemble those he was familiar with at Bluff Creek and did not accord at all well with the boys' estimates of its size. The front of the tracks was as big as anything he had seen before, with five large toes spanning eight inches, but the rest of the foot narrowed quickly and it had a total length of only 11½ inches. The tracks sank an inch into the ground in places where his own tracks did not show at all. Going up the hill there were tracks as deep as 14 inches in wet ground where Bob, by jumping down, could only make his heels dig in two or three inches. In the meadow where the thing was first seen he found a bed area 12 feet in diameter, which had a very strong smell to it.

When he challenged the boys of their height estimate of 14 feet they demonstrated that they could judge the heights of other things in the same range with consistent accuracy.

That was our first experience with a track that varied substantially from the Bluff Creek-Ruby Creek pattern, and although many strange tracks have been photographed and cast over the years there has never been another quite like those at Tenmile.

In 1963 I was back in California to see the tracks at Hyampom and I also had my first look at tracks in Washington.

In July, 1963, Mr. and Mrs. Martin Hennrich, of Portland, fishing on the Lewis River near its junction with the Columbia, saw a brown thing bigger than a human standing on the river bank. As they watched it walked into the trees. Later Jimmy Erion, who lived nearby, found 16-inch tracks a little farther down the river. I was told about those tracks by John Fuhrmann, a Portland postman and musician who also finds time to be a one-man clipping bureau, keeping a remarkable set of scrapbooks on a diversity of subjects. Although the sasquatch are just one of his many interests he has proved a very reliable correspondent over the years. Also in Portland at that time was Chuck Edmonds, now of Ashland, Oregon, a college art teacher who had done a very thorough job of research on several sasquatch incidents in Washington, Oregon and California. They showed me the tracks, which were by then several weeks old, and I made a cast of one. Most intriguing thing about them was that they came up out of and returned to the river, and I was told that when new they could be seen well down below the surface of the water as well as on the bank.

Chuck had gone to Fort Bragg, California, on the coast not far north of San Francisco, to look into a report that a huge hair-covered creature had tried to enter a house one night in June, 1962. He had taped an interview with Bud Jenkins, the owner of the house, which is one of the most interesting I have ever heard, as well as one of the most thorough. Here is part of Mr. Jenkins' story:

My brother-in-law heard the dogs barking and he got up and went out to see what they were barking about. We have a fence between the house and the barnyard made of six-foot pickets and he saw something standing by that fence looking over the fence towards the dogs, which he thought was a bear, and he came back in and woke me up and told me to come out and he would show me the largest bear I would ever see. It was standing upright.

I got up and went out with him and we didn't see it, so I said, "Well, wait a minute and I'll go in the house and get a flashlight and a gun."

So I came back in the house and my brother-in-law walked to the other side of the house then, to look back in the back yard, and as he stepped out from the corner of the house to look back there this creature stepped over this little two-foot fence we have out here right towards him, and he let out a scream and stepped backwards and as he stepped backwards he fell, so he came into the house on his hands and knees, going like mad.

My wife was at this time holding the screen door open for him to come in. I heard the commotion and I ran to the inside door which we have here before you step onto the porch, and as he came through the door I saw this large creature going by the window, but I could see neither its lower body nor its head, all I could see was the upper part of its body through the window here.

When he came in my wife tried to close the door and they got it within about two to four inches of closing and they couldn't close it. Something was holding it open. My wife hollered at me and said, "Hurry and get the gun, it's coming through the door."

Of course by that time I was standing right behind her in this door leading onto the porch, and I said, "Well, let it through and I'll get it."

At that time the pressure went off the door and she pushed the door to and threw the lock on it, and I walked to the window and put my hand up to the window and looked out, so that I could see into the yard, because it was still dark, and it was raining, and this creature was standing upright, and I would judge it to be about eight feet tall and it walked away from the house, back out to this little fence we have, and stepped over the little fence and walked past my car and out towards the main road....

I would judge it to weigh about 400 pounds and it walked upright at all times that I saw. It never went down on all fours at all, it stayed upright, and it had a very bad odor. The odor lingered on here for minutes after the creature was gone. And it left a hand print there by the door on the side of the house which was eleven and a half inches from the base of the palm to the end of the finger. It didn't act harmful really. It acted more curious than harmful, but it certainly gave us quite a start....

It was twenty minutes before my brother-in-law could hold a cup of coffee in his hand steady enough to drink it. Of course he stepped right to the creature and met it right face to face.

Chuck had also investigated repeated sightings at Dever-Conser, Oregon, in 1959 and 1960 of a white upright creature about seven feet tall. On one occasion it was reported to have run beside a moving truck looking in at the driver.

Another active investigator in that period was Lee Trippett, of Eugene, Oregon, along with his father, Ben. In October, 1964, Lee published and circulated a paper listing about 30 reports evenly divided between footprints and sightings, about half of which were new to me. He was also in touch with other investigators, some of whom I had not been aware of. Betty Allen had also published a little book about the Bluff Creek reports, and a year later, adding information from those two publications to what I had from other sources, I found that I had 120 reports in all. That was after nearly eight years of involvement with the investigation. Nowadays more reports than that surface every year, but they come from a much wider area.

Lee Trippett introduced me to Don Hunter, head of the Audio-visual Department at the University of Oregon. Here is the text of an interview that Lee recorded with Mr. Hunter in 1963:

My wife, who passed away in 1951, and I used to go up into the Three Sisters-Cascades area every year for vacation. This year, 1942, we went up quite late in the summer as was often the case. The war was on and there were not many people around because of travel restrictions and permits needed to get in the area. They were afraid of fire and sabotage. In this particular area we did not need permission. We had been, I think, at Big Lake and spent the night there.

The next day we went up to the mountain area exploring, taking pictures and the like. We came to Todd Lake, which is a little way north of the Century Drive, just before you come to the Sparks Lake. There was a forest camp there, but at that time it was not very well used and it was quite primitive. We arrived there around 3:30 or 4:00 in the afternoon.

As we got there I think we got out of the car and started looking around for a place to make our camp, when a rain storm came up. We sat in the car to weather out this storm as it was raining pretty hard. I looked out across Todd Lake. There is a meadow across the lake, and right in the middle of the meadow area there was a tall figure. It was just standing there. I pointed it out to Delores, my wife, and said, "My goodness, what's that?" It was difficult to see clearly from inside the car so we got out to get a better look. In getting out of the car we must have been heard because it took off for the trees with giant strides. I thought it could have been an elk or something seen head on, but there it was striding with two legs. There were no four legs about it. When he disappeared into the woods we were petrified. We thought it might come toward us, so we got back in the car and took off. We drove eight or ten miles away and spent the night in the car. It continued raining most of the night.

....When I had reported this to the Sisters Ranger Station I asked if there were any reports of a tall man in the area. I remember coming away from there rather discouraged. They seemed to think I might be nuts.

Lee also taped an interview with Gary Joanis, of Bend, Oregon, who told of shooting a sasquatch at Wanoga Butte in the fall of 1957. Like the boy at Tenmile he was using a 30.06 but the bullets did not down the animal. He said he had shot a deer and was waiting for a minute to be sure it was dead before going up to it, when a giant humanlike creature suddenly stepped into the clearing, picked up the deer and ran off with it. He emptied his gun at the animal's back, so close that he did not think there was any possibility he had failed to hit it, but it just kept going.

Another story that I learned from Lee was that of O.R. Edwards and Bill Cole, who reported a most unusual encounter while hunting together on Mount Ashland at the Oregon-California border in 1942.

Their experience shocked them so much that they never discussed it, even on the day that it happened, but 20 years later, after seeing published reports that other people had encountered the same sort of animal, Mr. Edwards wrote to Mr. Cole, who by then lived in Nebraska. Their correspondence led to Mr. Edwards putting the full story, as he remembered it, into a letter. Here is part of what he wrote:

It was just about sun-up now, just slightly hazy but the sun was burning it away fast and visibility was practically unlimited. The

THE QUIET YEARS

valley opened out....with a flat floor about a mile long and at least a hundred yards wide....This flat valley floor was covered with slick-leaf brush about shoulder to head high for the most part....The right hand slope was clear of any big trees, but there were numerous patches of brush....We sized up the terrain and I said, "Shall we try the brush patches up there?" and you said, "As good as any," so we angled up the hill to our right for 200 yards or so and came to the lower end of this particular brush patch. It was a good six feet tall, I couldn't quite see over it, oval-shaped, maybe 25 feet at the widest and 36 feet long up and down the slope.

You were to my left, so with just a nod by each of us you went left and I went right around the brush. We were both moving slowly and quietly. I was sweeping the area ahead with my eyes. On one sweep I caught a glimpse of what seemed like an apelike head just above the brush at the upper end of this patch. By the time I got my eyes back to focus on the spot it was gone. Then I heard the "pad pad pad" of running feet, heard the "whump" and a grunt as your bodies came together.

Dashing back to the end of the brush I saw a large manlike creature covered with brown hair, about seven feet tall. It was carrying in its arms what looked like a man. I could only see legs and shoes, straight down the hill on the run. I was about 30 feet away and the opening in the brush was only 10 to 15 feet wide. At the speed he was going it did not leave me much time to make observations.

I, of course, did not believe what I had seen, so I closed my eyes and shook my head to sort of clear things up, and looked down the hill again in time to see the back of the shoulders and the head of a manlike thing covered with brown hair disappearing into the brush some 70 or 80 yards below.

Bill, I was stunned. Basically I was okay. I checked myself over but I certainly did not believe what I had just seen. I went to the other end of the thicket where I thought I had seen something at first, found fresh scraps of leaves on the ground as if something had been pulling them off and eating them. The dry ground under the short grass was dusty as if it had been trampled, but I could make out no tracks that I could recognize. Perhaps if I had known what I was looking for I could have.

I walked on around the brush to where you should have been. I saw those dusty scuff marks and nothing more, stood there real still for quite a while, then turned and went up near the top of the ridge, found a little outcropping of rock and sat down on it. I was in plain sight from below and had a good view of the area where you had just disappeared. I lit a cigarette and watched and listened.

There is no need to say that I knew that something was wrong. Plenty was wrong. My hunting partner had just disappeared and there was no logical explanation. What I had just seen I did not believe. I waited and watched, lit another cigarette, then something else began to bother me. It was too quiet.

I sat there and smoked for over half an hour. Now, Bill, it seemed to me that this thing either packed you for quite a ways or you were out for quite a while. I guess I should have started looking for you. I don't know why, but I didn't. Maybe I was afraid of the creature, and maybe I was afraid I'd find you dead. Anyway I climbed up to the ridge. I followed it up to the head of the valley....

Half convinced that he was imagining the whole thing, Mr. Edwards continued hunting, circling the valley and returning on the opposite side where there was little brush and just a few scattered trees. As he neared the mouth of the valley he started angling downhill:

Approaching at this angle, I reached the edge of the brush-covered valley floor with about 200 yards more to go to the end. While the brush in the center of the valley seemed quite dense, the stream of water ran through there of course, there were clumps along the edge with room to walk between them. I proceeded to do this as it was much easier than following the steep slope. Suddenly I realized that I was following a beaten path. The grass here was taller, about eight inches, and dense, and had been pressed down to form a soft, noiseless path winding through the clumps of brush. The grass was not cut with hooves, so it was not a deer, sheep, cow or horse path. Something with a big soft foot, I thought, probably a bear.

I had nearly reached the end of the valley and the last clump of brush was about 30 feet away when from it came a very human "Shhht?" At the sound I froze for about a minute but nothing happened. I proceeded another few steps to within about 20 feet of the brush, and peering hard could make out a dark object or objects in the center of the small clump. The outline was so much like two men in dark clothing sitting close together that I spoke out with something like:

"Okay, this has gone far enough, I have a loaded rifle trained on you and I don't

THE QUIET YEARS

want to hurt anyone."

Not a sound, not a movement. Could be a black, burnt stump, but I had never seen anything quite like it. I moved slowly within ten feet and bent over to see into the brush patch. When I did this the right half of the stump moved with great speed toward the pine tree on the slope to my right and about 15 feet away. I thought I caught the outline of a great, long-legged man through the brush, but by the time I had jumped back to see around the brush it had disappeared behind the tree.

I stood still for quite a while hoping that old bear would come out from behind the tree so I could get a shot. All was quiet. Just a puff of dust from behind the tree drifting slowly in the quiet air. I then moved to the right around the brush. Four or five paces brought me almost directly between the clump of brush and the tree that something had disappeared behind. I bent down to peer into the brush from this angle and the other part of the stump went out the far side of the brush clump and, running bent over, back into the brush patch approximately the way I had come. I could not see the head or feet, but I could see the back and shoulders, which were flat and broad like a man's, and the rocking motion was exactly that of a man running bent over at the waist.

I would like to interject here that this second half of the stump had been definitely shorter than the one that went out first, and as I saw it go out the back side of the brush it seemed to be clasping something tightly to its chest with both arms. A young one, perhaps. I'll never know. I did get a good look at the back. Going into the brush bent over, the bending had opened up cracks in the fur so that I could see through the tips of the outside of rather light brown. The fur or hair underneath was quite dark like chocolate.

Since I was hunting I had instinctively brought my rifle to my shoulders as I watched, but since I was not sure just what it was I did not draw a bead. Then came the damndest whistling scream that I have ever heard, from right behind me. My hackles went up and I whirled to face the tree. Just in time to see a flash of something brown disappear behind the tree about six feet from the ground. More dust drifted from behind the tree.

I was less than 12 feet from this tree, and it wasn't any squirrel cutting off pine cones. I stood still covering the tree with my rifle. A full minute or more passed. Nothing happened. Then I moved a few feet to the south, past the tree, putting myself between those two things and the only exit I knew of, but still within 15 feet of the tree. Still covering the tree I stood real still for several minutes, probably four or five, straining my senses.

Then I noticed a knot on the side of the tree about six feet from the ground and to my right. As I looked there seemed to be an eye in the middle of this knot. I had been staring so hard and long that my eyes were beginning to water, so I shut my eyes real tight for a minute to clear out the tears, and when I looked again the knot was gone.

By this time I had had enough. I started walking slowly down the path or trail towards the car, looking back at every step. As I looked back at a distance of 60 or 70 feet I saw the head, part of the back and one outstretched arm disappear behind the brush, as if it had made a flying leap from behind the tree to the cover of the brush. I also heard plainly the "flop flop" of two feet landing on uneven ground at the end of the leap.

That was all, Bill. A very few minutes later I was back in your car, where you were waiting. You asked if I had seen anything unusual and I said, "Noooo, did you?" You said, "Oh no."

Then you asked me if I'd heard a scream, and on my negative answer you said you thought you had, but were not sure.

Bill Cole's recollection of what happened was not quite the same. He didn't think that the animal had carried him, although it had run into him, and he didn't think he had been unconscious. After the encounter he had abandoned hunting and returned to the car.

59

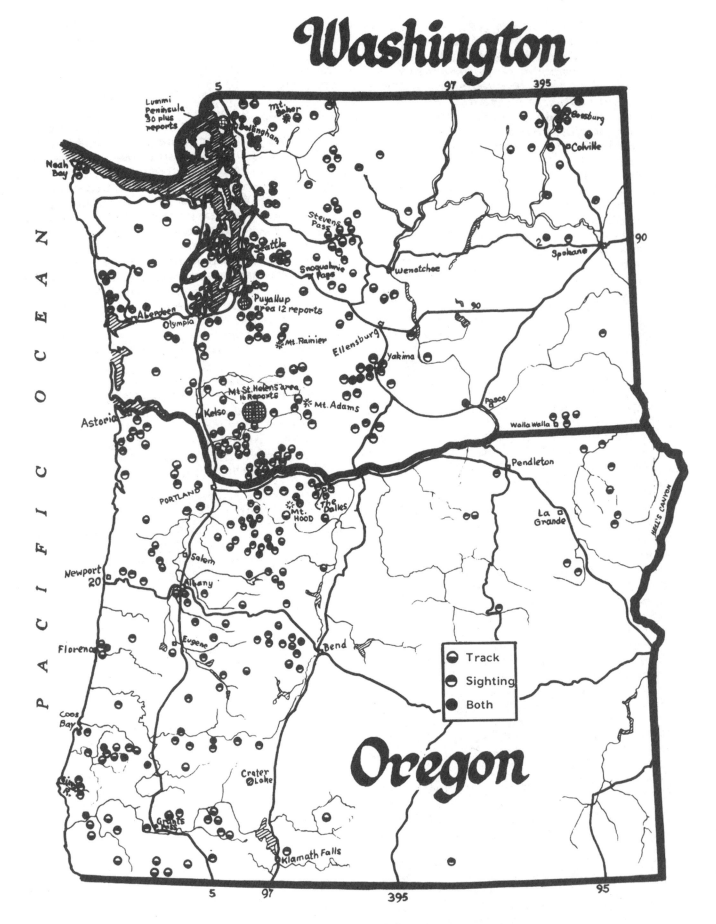

SIGHTING AND TRACK REPORTS IN WASHINGTON AND OREGON

Ape Canyon

Another person who began looking into sasquatch reports in the mid 1960's was Roger Patterson, from Yakima, Washington.

His interest in "Bigfoot" started with an article by Ivan Sanderson in *True* in December, 1959, but he did not do very much about it until 1964, when he went to Willow Creek and talked to a number of people there who were interested in the big tracks. Although it is about fifty miles away from the area where most of the tracks have been seen, Willow Creek is the nearest town of any size and has seen a lot of Bigfoot hunters come and go.

From there he went to Bluff Creek itself and had the good fortune to meet Pat Graves, a road contractor at that time employed by the Forest Service in a job that took him over many miles of dirt roads in the area and also out beyond the present roads into the bush.

Pat not only told him of numerous tracks he had seen over the years, some many miles from any road, he also was able to direct him to a set of tracks made just a few days before, and Roger was able to get an excellent cast of the 16" "Bigfoot" track.

That Roger saw tracks on a casual visit to the Bluff Creek area is by no means as coincidental as it might seem. There had been tracks to be seen on three of my trips to California when I had no advance information about them, and I had only been there twice when there were no tracks.

Unfortunately it has not been like that for more than a decade now.

While Roger was at Willow Creek he was able to see a collection of clippings kept by Betty Allen, and to talk to Al Hodgson, another local Bigfoot enthusiast. They gave him the names of numerous people up and down the country and got him thoroughly launched on his investigation.

One of the stories Roger learned about in California was the same "Ape Canyon" incident that Rene and I had found reference to in the British Columbia archives. We had never followed it up, but Roger did.

It had happened in 1924, the same year as Albert Ostman's adventure, but unlike the Ostman story it had attracted plenty of attention at the time. The story broke in the Portland *Oregonian* on July 13, at the top of Page One, under the heading:

FIGHT WITH BIG APES
REPORTED BY MINERS

KELSO, Wash. July 12 (Special) The strangest story to come from the Cascade Mountains was brought in Kelso today by Marion Smith, his son Roy Smith, Fred Beck, Gabe Lefever and John Paterson, who encountered the fabled "mountain devils" or mountain gorillas of Mount St. Helens this week, shooting one of them and being attacked throughout the night by rock bombardment of the beasts.

The men had been prospecting a claim on the Muddy, a branch of the Lewis River about eight miles from Spirit Lake, 45 miles from Castle Rock. They declared that they saw four of the huge animals, which were about 7 feet tall, weighed about 400 pounds and walked erect. Smith and his companions declared that they had seen the tracks of the animals several times in the last six years and Indians have told of the "mountain devils" for 60 years, but none of the animals ever has been seen before.

Smith met with one of the animals and fired at it with a revolver, he said. Thursday Fred Beck, it is said, shot one, the body falling over a precipice. That night the animals bombarded the cabin where the men were stopping with showers of rocks, many of them large ones, knocking chunks out of the log cabin, according to the prospectors. Many of the rocks fell through a hole in the roof and two of the rocks struck Beck, one of them rendering him unconscious for nearly two hours.

The animals were said to have the appearance of huge gorillas. They are covered with long, black hair. Their ears are about four inches long and stick straight up. They have four toes, short and stubby. The tracks are 13 to 14 inches long. These tracks have been seen by forest rangers and prospectors for years.

The prospectors built a new cabin this year and it is believed it is close to a cave thought to be occupied by the animals. Mr. Smith believes he knows the location of the cave.

There were probably as many as three of the prospectors still alive in the 1960's, but the only one ready to talk about what had happened was Fred Beck. Roger taped the following interview with him in 1966:

We was mining in there and working on our claims and about two years before we had the contact with 'em. We seen tracks down on the Muddy. We didn't know what they were. We thought that probably some big Indians in there fishing, barefooted. But we heard...there's...we heard rumors... people've been up there, year before that, telling the same thing too, seen them big tracks up there. One fellow was telling about fishing. Had a string of fish and laid them

APE CANYON

*on the bank, and when he went to get his...
looked around...there was a great big, like
a man, hairy, great fella. Had his finger
tips and just smashing them along the rocks.
He come out of there faster than he went in.*

*And nobody believed him. But mind you,
my father-in-law that was with me, he believed him. He said that fella didn't lie,
'cause he knew him. He said that fella was
scared when he came out of there, 'cause
he wasn't no human.*

*Then when we seen them tracks there we
never thought so much about it. My father-
in-law was telling about these people he
knew, his experience years ago seeing them
tracks up there and one thing and another.
And the Indians—there was lots of Indians
around here them days—them old Indians
was afraid to go back in there to Spirit
Lake 'cause they was all afraid of the spirit.*

*When we went up there why we never ex-
pected nothing like that. We heard noises
there, whistling, noise like pounding on
their chest.*

Q–When was the first time you actually
seen one?

A–*Well, about two years after that, we
were seeing them. Went out to a spring...
there was a little spring a little ways from
our cabin, our mining cabin. We had to go
down into a gully to get water, and there's
a ridge. The old man always carried his rifle
with him when he went to the spring, because he said he seen tracks, awful sus-
picious tracks, and after that he always carried his rifle. We went down. I went with him,
went down to get some water, and we looked
up on the ridge there, oh about 100 yards
away and we seen this peeking out of this
tree, and I seen him first and I said," Marion
there's a...look at that one on that tree."*

*He looked there and he just up and bang,
bang, bang, I see the bark fly, and the
bullets were all in just the place, just like
that (gesturing?) right around the bark of
the tree, just you know how....*

Q–*Yeah.*

A–*So we seen him running down this ridge
then, and then he took a couple more shots
at him. Marion when he first shot I rushed
over there, it was hard going, he said:*

*"Don't run, Fred, don't run," he said,
he won't go far," he said, "I put three
shots through that fool's head, he won't go
far."*

*So we got up the ridge and looked down
there he was goin', just jumpin', looked like
it'd be twelve, fourteen feet at jump, runnin'.
The old man took a couple more shots at him
and the old man said:*

"My God, I don't understand it, I don't

**Fred Beck and the gun he used to shoot one
of the "Mountain Devils" at Ape Canyon.**

*understand it, how that fella can get away
with them slugs in his head," he says. "I
hit him with the other two shots, too."*

Well he got away alright.

Q–*Well now the tracks that you seen before,
you would estimate at around how big?*

A–*Well, say 19 inches they measured. You
see the two detectives up with us from Portland and they measured the tracks. It rained
that night. We stayed at the lake. A regular
cloudburst. We went up there and we had
where we washed the dishes, right under
the little seep hole, you know, springs, you
might call them springs, melted snow, kind
of a little gully there. We had cleaned our
pots out, our beans and rice. It quit raining
before we got there. When we went up there
there were tracks made since it rained, fresh
tracks, about 19 inches long. And the detec-
tive said, "Well, these are tracks all right,"
but, he said, "who made them?"*

*I said, "Well, who did make them?" I said,
"We was all together, and they've been made
since the rain."*

*And they talked about it and they said it
might have been a coon making them, or part
human—all that stuff—a bear part human!*

Q–*When you finally did see....*

A–*When we seen 'em, you know, why we
heard that noise—pounding and whistling, at
night they come in there and we had a pile
of shakes piled up there, big shakes. Our
cabin was built out of logs. We didn't have
rafters on it, we had good-sized pine logs,
you know, for rafters, two-inch shakes,
pine shakes. We had them rafters close apart,
they was about a foot apart, 'cause he said
he wanted to make a roof what'd hold the
snow. We made one to hold the snow.*

APE CANYON

Them buggers attacked us, knocked the chinking out on my dad's, on my father-in-law's chest, and had an axe there, he grabbed the axe. And the old man grabbed the axe and turned it so it wouldn't go through the chinking hole between the logs and then he shot on it, right along the axe handle, and he let go of it. And then the fun started!

Well! I wanta tell you, pretty near all night long they were on that house, trying to get in, you know. We kept shootin'. Get up on the house we'd shoot up through the ceiling at them. Couldn't see them up there, you could hear them up there. My God, they made a noise. Sounded like a bunch of horses were running around there.

Next day, we'd find tracks, anywhere there was any sand on the rocks, we found tracks of them.

Q-Were those the same tracks that you'd seen?

A-Yes, the same tracks, them tracks, but never measured them. They were big fellas. That was the only track we measured, was them the detectives measured.

Q-Now, did you explain to us a while back that there was at one time that you had shot one, when did this....

A-Well that was the next morning, I guess it was, if I remember that.

Q-After the attack?

A-Yeah, I was, we was going to go down to the tunnel to get some tools out that we had, drills and things like that in the tunnel, we was going to clean out and go home. When we went down I took a rifle with me. We all carried rifles after that happened.

I, down the ridge there a couple of hundred yards, no it wasn't that far, why there was one of them fellas run out of a clump of brush and run down the gorge, and I shot him in the back, three shots, and I could hear the bullets hit him and I see the fur fly on his back. I shot for his heart. And he stopped and he just fell right over a precipice, and I heard him go doonk, zoop, down into the canyon.

Q-You said he fell into Ape Canyon.

A-Yeah, and the sun come out in the afternoon, that water was really a torrent goes down there, it'd wash anything out fall in there. And that's the reason I don't know if they're human or not, cause I couldn't kill 'em. And I hit.

Q-Well how would you describe, Mr. Beck, as far as what they look like in their body and their head?

A-Well, they was tall, I dunno, they looked to me like they was eight foot tall, maybe taller, and they was built like a man, little in the waist, and big shoulders on, and

chest, and their necks were kinda what they call bull necks, you know how they are.

Q-No neck at all, hardly.

A-That's it, and then their ears, turns out like ours do, and so big, you know, and hair all over, you couldn't tell nothin' about em.

Q-Did they have hair on their face, or could you, did you ever....

A-No, let's see, I don't believe they did ....I believe they did have hair on their face.

Q-But not as much as...

A-No (cough) can't have whiskers.

Q-Sure. How about their nose?

A-I couldn't of but I was, uh, they seemed to have a kind of pug nose, flat nose, kind of flat.

Q-And their eyes?

A-All I can know is, we were excited, you know, you don't see very good detail when you'r excited, but I know one thing, that they was no human.

Q-They did, though, walk upright. Did you ever see...

A-I never seen one on the four.

Q-Their arms, probably, was they...

A-Arms down below the hips, long, I figured....

Q-Below their knees?

A-Yeah, their knees. Long arms. And big arms.

Q-What would you estimate maybe for weight?

A-Pretty heavy. I'd say they're six or eight hundred pounds. Like you know, estimated. And maybe more, I don't know. I couldn't tell ya the weight of them. The way they sunk down in the ground I'd have some idea about it. I'd say they weighed eight, nine hundred pounds, or more, 'cause it made a deep imprint in the dirt. There's so much rock up there, you know, only can see them in places where there's sand, you know.

Q-Sure. Well now after you had the attack, what happened then, the next morning?

A-Well, we come out, out of there. Come down, and my father-in-law he was so excited and scared. I told him, he promised never to tell anybody, 'cause I said it wouldn't do, people wouldn't believe it, don't tell anybody. He say, "I won't, I won't," but he did. Went down to the lake and the rangers down there knew him. He was so excited they found, took him in the other room and talked to him an' he acknowledged what the trouble was. They said they believed him, because the old man had been a hunter, they knew him. All his life...hunting until no little thing would ever scare him, no animal or anything like that. Then he

APE CANYON

went to Kelso and told some of his friends down there. Then the newspaper reporter give us a merry time, day and night.

Q-Had they ever heard of anything, anybody before this?

A-I don't think so. They really didn't believe it.

In my own discussion with Fred Beck he explained more fully the incident of the chinking and the axe. The spaces between the logs in the cabin were wide, and large strips of split saplings were wedged in to close the holes. One of these was knocked out above the berth where Marion Smith was lying and it hit him on the chest. Then a hairy hand and arm came through the hole, groped around, and got hold of an axe by the handle. As the axe was disappearing through the gap in the log wall, Smith managed to turn the head at right angles to the crack so that it could not be pulled through. Then he grabbed a gun and shot down the line of the handle, at which the axe was released.

As far as I could determine, that hairy arm was the only part of any of the attackers that the men saw that night. Their cabin had no windows. There must have been at least two creatures outside, as there was one heard on the roof at the same time that one was heard pounding on a wall. The impression is that there were more than two, but that could not be definitely established. Mr. Beck said nothing about a barrage of rocks coming into the cabin. He did mention hearing rocks land on the roof and roll off. At that time I had not seen the newspaper reports and did not know that he was reported to have been knocked out by a rock.

I got the impression that Fred Beck had told his story so often that he had established a set pattern of things to say and there wasn't much use asking further questions. To my understanding there was a difficulty in fitting all the elements of his story in logical order, but I was not able to clear that up.

Did all this really happen? I think so. To the people at that time and place, knowing nothing of such creatures except the old legends of mountain devils, the miners' story was not believable. However if such animals do exist, then certainly the most acceptable explanation for the miners having claimed to see them is that they did see them. There isn't a shadow of a suggestion as to why they would make up such a story and keep telling it all their lives.

The top end of Ape Canyon. It widens out farther down the mountain.

Something On The Road

By 1966 Roger Patterson had attracted a fair amount of publicity for his sasquatch investigations and he began to learn of reports in the vicinity of his home town of Yakima and also in the "Tri-Cities" area of Pasco, Kennewick and Richland, where the Yakima River joins the Columbia. Nearly all these reports concerned a white creature reportedly over seven feet tall, and they were very recent.

Close to Richland, and far from any mountains, this creature had frequented an unused gravel pit, where it was seen repeatedly by groups of teenagers, some of whom made a regular game of "hunting the white demon." Rene and I talked to several of these boys in 1967 and found that they sounded convincing. We also learned first hand that at least some of their parents had known what was going on at the time, but neither they nor the authorities did anything about it. We found it hard to understand how groups of teenagers could be allowed to cruise the countryside in the small hours of the morning armed with rifles and shotguns and shooting at a "monster," without somebody doing something about it, but that seems to have been the case. The dry, open country did not seem at all suitable for a sasquatch, but there seemed no doubt that something had been there.

At Yakima, we were told, there was a youth who had been chased by this white creature in an orchard and had required hospital treatment for shock. However we did not meet him. We did interview another Yakima youth (who later became a deputy sheriff) and he impressed us very favorably indeed. His story, as recorded on tape, was as follows:

It was around the 19th or 20th of September (1966). I was on the Fisk Road about nine or ten miles west of Yakima about 11 o'clock on a Tuesday night and I had taken a friend of mine home that lives out there.

I was on my way home and I came round the corner and it was raining that night, and lightning flashes and everything, and this thing was standing in the middle of the road. I slammed on the brakes and my car skidded right up to it and stopped about three feet from it. It just stood there and looked at me right through the windshield of the car.

It killed the engine on my car when I slammed on the brakes. He walked around behind my car and then turned around and came right back to the window and just bent over and looked in the window at me. I got my car started and took off.

Q-What did it look like?
A-I imagine it was a bit over seven feet tall, pretty good size, had long hair all over. It was just in the form of a big man.
Q-Would it be a heavily-built man?
A-Oh, yes, awful heavy. He was awfully broad.
Q-What colour was he?
A-Kind of a light greyish white colour.
Q-How could you judge the height of it?
A-By my car. He towered over my car by a long way.
Q-When he was looking at you I guess you got a look at his face. Could you describe that?
A-He had a real flat nose and his lips were real thin. He had hair on his face too, there wasn't any skin other than his lips and just around his eyes that I could see. I couldn't see any ears. He had long hair on him, all over his chest and everything.
Q-What colour were his eyes?
A-Well when the car lights hit him his eyes kind of had a red...just a...almost flourescent, like a rabbit's eyes or something when you shine a light in them at night.
Q-What did they look like when he was looking in the car?
A-Just eyes....I was so busy trying to get my car going I wasn't concentrating on staring him in the eye.
Q-Did he seem to be at all menacing?
A-No, it didn't try to touch my car at all. It never made any noise whatsoever. Just looked through the window.
Q-Were his lips open?
A-Well kind of, I could see his teeth. His teeth weren't very big in front except these (pointing out eye teeth) and they kind of protruded more than the others. They weren't fangs, they were slanted more.... forward. And his face was slanted an awful lot. He had a real low forehead.
Q-Did he have a big jaw?
A-Oh yeah, a great big square jaw.
Q-Did the nose protrude like a human nose?
A-Well not very much. It was real flat and wide.
Q-As if a human nose was squashed?
A-Mm hm.
Q-Did you see him walk?
A-Well just back to the back of my car. He only took about two steps, just plunk, plunk and that was it. He took a stride about half the length of my car.

The car finally started and he left the sasquatch standing in the middle of the road. Roger heard about the incident soon after it happened and was able to see the footprints left in hard dirt beside the road, where a man "didn't even make a mark."

SOMETHING ON THE ROAD

At about the same time Roger was invest-igating that story in Washington I was re-cording one in British Columbia that was a good deal less tense for the driver involved but was almost its twin in the detailed des-cription of the animal. The story was told by a bus driver. It had happened a few year before, in 1962 or 1963, while he was driving home in his own car after midnight, between Vedder Crossing and Yarrow, some 60 miles east of Vancouver. It was pouring rain:

I was going quite slow because of the severe curves in the road and as I came around the final curve I saw what I thought was an extremely large man standing along the edge of the road. And then as I got close to him I realized it wasn't a man or anything I had ever seen that looked like one, and when he looked at me my headlights hit his eyes and they glared. Then he slowly kind of half walked and half loped across the road, and I kind of turned my car and fol-lowed him with my headlights. Then he stop-ped along the edge of the road and looked at me for a second and then jumped onto the bank.

Q-How tall would this thing be?
A-I would say he was well over seven feet.
Q-Any clothes?
Q-No, none that I could see whatever, just quite long hair all over his body.
Q-Would that be similar to an animal's hair?
A-Yes, it looked very similar to that, like a large ape, hairwise, except much taller.
Q-How close did you get to him?
A-Twenty-five feet, twenty to twenty-five feet.
Q-How fast would you say you were dri-ving?
A-I was going quite slow, no more than 30 miles an hour.
Q-Did you continue at that speed?
A-No, I slowed right down. Actually I stopped completely.

Q-You stopped and watched him?
A-Yes, until he disappeared.
Q-You say his eyes shone?
A-Yes, they had an animal glow to them
Q-Did you notice any particular colour?
A-No, I would say it was a kind of a red-dish kind of glow. It would be hard to des-cribe exactly. At that moment I was kind of excited and I didn't get too good a look at his eyes.
Q-Did he look heavy?
A-Yes, I would say he was well built, very athletic-type build actually, and very well proportioned from what I could see of him. I imagine he weighed well over 400.

Q-Was he watching you all the time?
A-Not at all times, he was looking up the hill...actually he only looked back directly at me twice.
Q-He wasn't really paying that much atten-tion to you then?
A-No, he wasn't.
Q-Were you ever at any time able to see any of his features, any part of his face?
A-Yes, I saw part of his face when I was right close to him—a very apelike, flat-nosed some of it quite hairy and very thin-lined mouth as far as I could see. The one out-standing feature is, I didn't notice any ears. The only other thing was the lack of neck. He had little or no neck at all.

He did not hear the animal make any noise. When it jumped up the bank it just took off without apparent effort and disappeared above the area lighted by his headlights. After a while he cautiously marked the spot by poking a foot out of the door of his car and scraping some rocks together with his toe. Next day he found that the thing had to have risen six feet on its first jump, to land on a large rock at the bottom of a slide.

The most usual way for anyone to see a wild animal nowadays is to encounter it on the road while driving, and it is the same with the sasquatch. Those two reports were by no means the first we had learned of, but they were the first to involve really detailed observation. Jumping ahead a bit with the story, there are three more such reports that are worth quoting in full from the tape-recorded interviews. The first was a sighting near Hoquiam, Washington, in July, 1969, by Verlin Herrington, at that time a deputy sheriff in Gray's Harbor County. With the co-operation of the sheriff I was able to go over the incident with him a few days after it happened. The following interview was taped at the spot where the sighting took place.

Q-Now the incident we've been discussing, could you give us the date when that took place?
A-Yes it was July 26, on a Sunday.
Q-And the year?
A-1969.
Q-And what time?
A-It was 2.35 a.m.
Q-What were you doing at that time?
A-I had been on an incident at Humptulips and I was en route by way of Deekay Road, by Grass Creek Road, to the beach and into my residence.
Q-You were working at that time?
A-Right.
Q-What was it happened while you were making that trip?

66

SOMETHING ON THE ROAD

A-As I was going down Deekay Road I rounded a corner, and my first impression was of a large bear standing in the middle of the road. I either had to stop for the bear or hit him, so I decided to stop, put on the brakes, came to a screeching halt and coasted up the slight grade as far as I could without startling the animal as I was looking at it as I was going towards it. This animal in my opinion was not a bear, because you could see by the way it was standing that it had no snout. It had a face on it. Its eyes reflected, and when I came to a complete stop I could see in the headlights of my car that it had feet on it instead of paws, and it had breasts. When I centered my spotlight on the patrol car on it, it walked to the edge of the road. It didn't fall down on all fours like a bear would. It walked to the edge of the roadway and stopped, turned, still looking at me. I re-adjusted the spotlight on my car so I could look at it better. Its feet had hair down to the soles but you could see the outline of the foot. It did have toes. Its hand was in a position where it was spread out, and it did have fingers. After I'd re-adjusted the spotlight, I rolled the window down, pulled my revolver and crawled out of the door. I aimed my revolver at it and as I cocked the hammer on it the animal went into the brush. I got back in my car and drove off.

Q-How far away were you when you were looking at it?

A-Seventy five to eighty feet.

Q-It was standing erect on its hind legs?

A-Yes it was. As it went off into the brush it was still on its hind legs.

Q-Where it went into the brush, that's down quite a steep little bank isn't it?

A-Yes, it is.

Q-And could you see that it remained erect as it went down the bank?

A-Until it went out of the spotlight, yes.

Q-Just to sort of go over it from top to bottom, can you say anything about the shape of its head? For instance your first impression was a bear. Did it have a snout?

A-No, it didn't have a snout. I couldn't say it has a nose like a person would have. I believe that there was no hair on its face. It had a dark leathery look.

Q-Did you get any impression of the length of hair on the head or on the rest of it?

A-I would say about three to four inches on the head.

Q-It didn't have long hanging hair at all?

A-Longer on the head than on the body, yes.

Q-What sort of neck did it have?

A-No neck.

Q-You mentioned, I think, that it had breasts.

A-Yes.

Q-Where on the body were they located?

A-Like a human person's. They were also covered with hair except for the nipples, and they were skin.

Q-Did you mention the color of the hair?

A-No, I didn't. It was brownish black, dark colored.

Q-You say you saw a hand with the fingers extended?

A-Yes.

Q-Could you tell how long the arms were?

A-I would have to guess at how long they are. The arm that I was looking at was in a bent position like you catch someone startled.

Q-You couldn't at any time see how far down the body the arm would come?

A-No.

Q-Comparing it to a human, would you say that it had long legs or short legs?

A-Long legs, I would say, long muscular legs.

Q-Did it take long strides?

A-No, it seemed like it took small steps as it walked. It was watching me as it walked to the edge of the road. They would be large steps for a human, but for this animal they were short steps. About three steps from the center of the road to the edge.

Q-Did you notice any difference in the way it walked and the way a human would walk?

A-Same type of position and same type of walk as a human being.

Q-Did it taper at the waist? Did it have a narrow waist?

A-It was a big body, a big stout body, that's the best I remember it was. It could have been thinner at the waist, I don't remember.

Q-It was pretty heavily built?

A-Yes.

That report of a sighting by a deputy sheriff on duty was widely publicized, although an attempt was made to change the story and pretend that he had really seen a bear after all, and it led to several other reports of sightings out on the Washington coast being brought to light. Like the two preceding reports Verlin Herrington's sighting was at night, but the following January I learned of a close-up sighting on a highway in broad daylight.

The witness was the foreman of the road maintenance crew at Squamish, British Columbia. He reported the occurence to his superintendent, who passed the word on to the provincial museum. I was notified by the museum staff and met the foreman at the site a few days later. This is what he had to say:

67

SOMETHING ON THE ROAD

I came around a slight curve in the road and there was this animal on all fours facing away from me, a large hairy animal. I thought at first it was a bear. As I drew closer it turned sideways and it got up on its hind legs and ran across the road in an upright position. I knew then it was definitely not a bear, it was a human like animal but a monkey type appearance.

Q-How big did it appear to be?

A-It was seven feet tall, at least seven feet and 250 pounds...very prominent stomach.

What color was it?

It was reddish brown pretty well all over the body from the shoulders down, the head was a darker brown hair and not quite as long as the rest of the body. On the head was not much more than a medium type crew cut. It was not flat at the top or anything, it was all the same length, like the back of its head and the top of its head it was a rounded type haircut, and on the body was very shaggy, about four inches long, like a bear.

Q-You said it was a monkey type thing, just what were you referring to?

A-Well, just the general appearance of the thing, a large, large monkey, and definitely not a bear. It didn't have a bear's face, it had more of a flat face like a person or a monkey. Its face was much lighter colored than its body hair, and definitely lighter than the head hair and it appeared to be almost hairless I would say but a lighter color.

Q-How far away would you be?

A-When I first noticed it I would say probably two to three hundred feet and when I slowed down to finally a complete stop to have a look at this thing that I thought was a bear...I wanted to have a good look at it ...I would say approximately 100 feet when I stopped to watch the animal.

Q-How long do you think it stayed before it started across the road?

A-Not long at all, it either fell off this bank or it fell on the shoulder of the road and picked itself up immediately and just scurried right across the road and up the bank here on the other side, in the upright position like a man would run, and it did run, it didn't waste any time at all.

Q-Did it move very fast?

A-Quite quickly. Also it looked at me when it was standing up and pretty well running all the way across the road, but the first impression I got was it was either terrified or it was very angry, the expression on its face, and then I was terrified, actually, because it was something you don't

expect to see.

Q-Could you see the arms or forelimbs distinctly?

A-Yes, it definitely had arms...quite long arms. In proportion to its body I wouldn't say they were longer than a man's to his body. It definitely had hands. I'm not sure how many fingers but it did have hands and in the right hand it was carrying what appeared to be a fish, a 10-inch fish.

Q-How was it carrying it?

A-It had the hand wrapped around the fish and the head was on one side and a length of the back and the tail on the other side, the front and back of its hand. It had its fist closed around the fish.

Q-Could you see it well enough to be sure that's what it was?

A-I would definitely say that that is what the animal was carrying.

Q-Did it use its hands at all as it was moving?

A-Its hands swung along when it was running across the road, and when it was climbing the bank it did use its left hand a couple of times, it bent slightly forward like a man would do and assisted itself or balanced itself up the incline with its left hand, still holding the fish in its right hand.

Q-How did the length of the legs seem in proportion to the rest of the body?

A-They were about comparable to a man's legs. Its legs were not overly long or short for the size of the animal. It did have a prominent stomach, that is what I did notice.

Q-Were the limbs slim or heavily built?

A-Very sturdy type build, but hairy as well, so there would be no way to tell whether it had terrific muscles, but sturdy, a very sturdy animal.

Q-What was the neck like?

A-It didn't have a great deal of neck at all, the head just kind of sat on the shoulders.

Q-Could you tell if it was male or female?

A-There wasn't any sign, or at least there was no evidence of sex that I noticed.

Following the taping of the interview we tried re-enacting the incident, with the foreman in his pickup where he had stopped before, while I ran across the road at the point where the animal had been. He then had second thoughts about his estimate of size. I am over six feet and he said the animal looked half again as big as I did. That would also indicate a much more substantial sized fish. Nearest water is a river running far down below the road in a steep canyon, so if the creature was indeed carrying a fish it seems probable that it was not for its own consumption.

There are a number of people who claim

SOMETHING ON THE ROAD

not only to have seen a sasquatch on the road but to have run into one. I have never met one of these people but I do have a copy of a tape on which a truck driver tells of such a collision in mid-January, 1973:

I was hauling a load of logs out of Grant's Pass, Oregon, to Eureka. At that time we had to go down the Avenue of Giants. It was about 7:30 in the evening. I was on the highway by myself so I had all my lights on. I was doing about 40–45 miles an hour when I went around a curve and as I got approximately to the center of the curve this being, whatever it was, stepped out from the right hand side and I hit it with the truck. The top of the hood is six foot four inches high and the top of his head was about six inches above the top of the hood. I don't know why it didn't see the lights or notice me. I didn't notice whether it was male or female. The upper torso was turned away from me at possible 45 degrees rotation.

I hit it with the truck and it flew off to the left hand shoulder of the road. I didn't hear any noise it made. Of course I couldn't hear anything over the exhaust.

After I hit it I drove down the road about five or six miles and then stopped to check on the damage, which was quite extensive. I didn't see any blood or hair on the front end. The front radiator shell had been smashed in, away from the fibreglass hood. The glass was cracked in several different places, and we had to replace the whole thing.

Q–You say you have had trouble with bears, you've hit bears before?

A–No, I've hunted bears before. So I know it wasn't a bear, I've hunted bear too often to make a mistake. If the body was in proportion with the torso the legs would have been too long for a bear. It did not walk like a bear, it walked almost erect. There was just a little bit of forward angle from the waist. The shoulders didn't look hunched at all. The head, from what I could see it didn't look like there was a neck at all, the head sat right on the shoulders.

Q–Did you notice the shape of the head?

A–From the back side it seemed to be round, not like I would picture a gorilla or an ape or something with a crown on the back of the head. It was more humanlike. The arms and hands were long enough to... they stopped about an inch above the knee.

Q–What color was it?

A–It was, I would call it auburn—almost a russet color.

The driver reported the incident to a supervisor, who asked what he had been drinking.

The author measuring the stride of the larger of the two tracks on the Blue Creek Mountain road.

Blue Creek Mountain

On the last Sunday in August, 1967, I returned from a trip to northern California with Harold McCullough and Dale Moffit from North American Guard Dog Service. I had seen a few poor tracks in two sizes, 15-inch and 13-inch, the first tracks I had seen for four years. Purpose of the trip had been to try a tracking dog on the tracks, but they were too old and trampled over.

While I was there I talked to Mrs. Bud Ryerson, who had seen the tracks when they were new, and I gave her my phone number with a request that she call me if more tracks were seen. I also stressed that it was important that they be uncontaminated by human scent.

On Monday morning at seven the telephone woke me. It was Bud Ryerson, whom I had not met, and all he said was that what I was looking for was there. He was calling on a radio phone, and he never mentioned "Bigfoot" or tracks. I asked him when and he said during the preceding evening. That sounded very promising, so I started phoning around to try to round up a dog, money and scientists to look at the tracks.

I drew a blank on the scientists, but to my surprise was able to get $500 from the Vancouver *Sun* to help pay for an airplane. There were difficulties and delays, but shortly after 1 p.m. Dale, Rene and the dog and I were packed into a Cessna 185, heading south.

We arrived at Orleans, near Bluff Creek, at dusk, and it was fully dark by the time we got to where the tracks were. The dog showed great interest, but none of us wanted to follow those tracks into the bush in the dark.

Mr. Ryerson showed us where the tracks had been noticed first, at a place where a few metal parts of a tractor and some smaller items in boxes had been left near the road. These had been scattered out onto the road during the preceding night, which had quickly drawn attention to the big prints in the deep dust。

Mr. Ryerson said that there were three sets of the big tracks wandering along the road, one larger and two smaller. We studied a few prints closely by flashlight. They were familiar to me—the same 15-inch print with a split in the ball of the foot that I had first seen nine years before, but I had never seen such sharp detail. The prints had been made while a thin layer of dust was damp from a brief rain, and this had dried, so that the impression did not crumble.

Mr. Ryerson had done all he could to keep the tracks undisturbed—even interfering with his own road-building operations, but he could not stop all traffic. Cars had already wiped out the tracks on the travelled part of the road and every vehicle that passed stirred up a heavy cloud of dust,

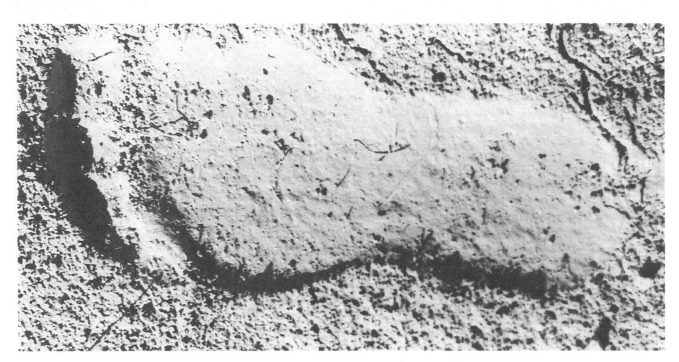

A typical print of the 15-inch track on the Blue Creek Mountain road.

70

All of these casts are from the tracks on Onion Mountain and Blue Creek Mountain, except the little one in the lower right corner, which is a bear track, and the one next to it, cast where the trailers were parked on Bluff Creek. The double cast, with both 15" and 13" tracks in one cast, is now at the B.C. Provincial Museum.

so that the tracks would obviously lose detail very quickly.

By this time we were beginning to realize that there were an awful lot of tracks. The following day we counted 590, and there must have been well over 1,000 in the beginning. They were on the road in two sections totalling about 600 yards, with a gap of several hundred yards where the makers had travelled to one side of the road.

The tracks were right at the summit of Blue Creek Mountain, which was actually just the high point on a fairly level ridge, but was close to 5,000 feet in altitude. Mr. Ryerson was contractor for a new access road being built along the ridge, replacing an old jeep road. The previous tracks had been about five miles farther south, on the same road.

We were back with the dog at daybreak, taking up the track where the 15-inch foot left the new road and turned down an old jeep road baked too hard to show a track. Immediately we were disappointed, for the dog showed none of the interest of the night before. Told to track, she wandered erratically, sniffing under the small bushes beside the road. Dale told us that was the last place where a trace of scent would hang. Eventually she took us a few hundred yards down the old road and then came back.

That finished the primary object of the trip, but in the meantime we learned that

the B.C. Museum was sending a anthropologist down to look at the tracks, so we decided to wait until he arrived. We took a lot of photographs and counted and studied tracks, but as the day wore on the heat went over 100 degrees. We had had very little sleep, having driven back to Orleans in the middle of the night to telephone, and we probably didn't accomplish half what we should have in studying the tracks. Nor did we make any systematic search for hair.

With two wide strips of road showing only tire marks we were no longer able to confirm that there had been two smaller tracks with the big one. There were many places where there were two sets of tracks, one on each side of the road and sometimes both on one side or one in the middle, but at the only place where there were three sets two of them were the large size. It appeared that the larger individual had wandered into the bush and then returned to the road farther back, covering the same stretch of road twice.

The smaller track left the road a considerable distance before the larger one, stepping up onto a bank about two feet high. Off the road the ground was rocky and overgrown with low, stiff brush, conditions where none of us could follow a trail.

Don Abbott, B.C. Provincial Anthropologist, arrived the following night. In the meantime we had made a few casts, but we

BLUE CREEK MOUNTAIN

were saving the clearest tracks for him to see, and we had checked the same two tracks on a sandbar down in Bluff Creek, about 10 miles away and several thousand feet lower down. They had been made the same night as those on the ridge, but we did not learn about them until two days later. There were very few tracks left, as they had circled some trailers where loggers were living, and when the loggers came back from a weekend in civilization and found what kind of visitors were in the area they removed the trailers. Most of the area was rutted with fresh tire tracks, but a few good footprints remained, showing the extreme depth that is always obvious in damp sand, where human prints hardly sink at all.

Don Abbott's approach to the matter of preserving tracks was a new one to me. He proposed to use glue and sacking to solidify the ground itself and lift the actual track instead of casting it. We spent much of one day working on two or three prints in this way.

Next day he contacted Humboldt State University at Arcata and persuaded several zoologists to come to see the tracks. This did not prove to be easy, he said, even though they were only eighty miles away. They treated him as a crackpot until he happened to mention that the tracks were those of a "bipedal primate." After that,

apparently convinced that he spoke their language, they decided to risk the trip.

Waiting for them to arrive proved to be very costly. We did not get back up the hill until mid-afternoon, and in the meantime it had been decided that we were not coming back and a grader operator wiped out almost all the tracks, including those we had so painstakingly tried to preserve the day before.

The Humboldt State men did not see enough to have anything to base an opinion on. We cast what few poor tracks were left, and Don cancelled a trip by Charles Guiguet, the curator of mammals, who was to have come down next day.

None of us, including Don Abbott, had any real doubts, while the tracks were there to look at, that they were indeed tracks and not a hoax. However Don is a cultural anthropologist, not a zoologist, and the zoologists lost no time in trying to persuade him to revise his opinion.

That was the first time, as far as I know, that a representative of any scientific institution was ever sent to study the big tracks anywhere in North America. It resulted in the museum director, Dr. Clifford Carl, and the minister of recreation and conservation, Hon. W.K. Kiernan, publicly requesting more information about the sasquatch. The response was not heavy, but it did add a few new reports to the file.

The 13-inch print in a sandbar. Compare the depth with that of the boot print at top right.

72

Roger Patterson's Movie

I first met Roger Patterson in 1965 when he was working on a book, published the following year with the title *Do Abominable Snowmen of America Really Exist?* He came to see me to get permission to use some of my material, and we spent a long evening together. Early in 1967 I met him again at his home near Yakima, when Rene Dahinden and I made a trip through Washington, Oregon and California to call on other people active in the sasquatch investigation and to make new contacts in northern California where nearly all of our old friends had moved away.

We called again on the return trip, and in all spent about two full days with him.

While in California we learned that Bluff Creek and vicinity had experienced very heavy rains in 1964 that had caused many land slides and serious flooding. Since then we were told, the big tracks had not been seen. Persistent enquiry finally turned up a total of three reports for the years 1965 and 1966. When I was back in California with the tracking dog in August, 1967 I notified several other "sasquatch hunters" that there were tracks there. Roger was among the people I tried to contact, but I could not reach him and he never saw the Blue Creek Mountain tracks. However he came to Bluff Creek a month or so later, prepared to stay for a while, hoping to get pictures of fresh tracks for a movie he hoped to make about his search for Bigfoot.

Roger and Bob Gimlin, also from Yakima, set up camp near the end of the road in the bottom of the Bluff Creek valley and began making daily patrols on horseback to inspect the sandbars along the creek. It was on one of these patrols that they rounded a bend in the creek and confronted a dark animal crouched by the water. The animal merely stood up and looked at them, but their horses shied and Roger's lost its footing as it reared and fell on its side. By the time he got himself disentangled from the horse and got his movie camera out of the saddle bag the creature had started to walk away and he ran across the creek after it, filming as he ran.

Most readers will have seen the movie by this time, as it is shown from time to time on television. It shows a female creature, just such as so many people have described, walking erect on two legs, extremely heavy of build, and completely hair-covered. The movie is not sharp enough to show facial detail but it is plain that most of the face is hair-covered, and that the head has heavy brow ridges and a peak at the back, like

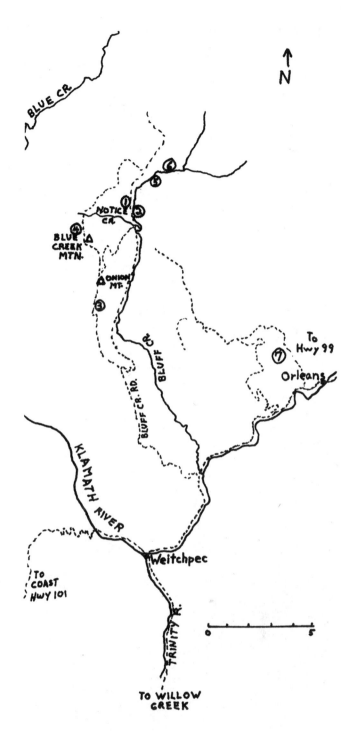

THE HEART OF BIGFOOT COUNTRY

(1) Jerry Crew's Bigfoot tracks.
(2) Bluff Creek sandbar, first 15" tracks.
(3) Onion Mountain tracks, August, 1967.
(4) Blue Creek Mountain tracks, August, 1967.
(5) 15-inch and 13-inch tracks on sandbar, August, 1967.
(6) Patterson movie and 14½-inch tracks, October, 1967.
(7) 14½-inch tracks seen in 1960.

Locations are all approximate.

73

ROGER PATTERSON'S MOVIE

a gorilla, but does not have an ape's projecting jaw. It has no neck at all, a point emphasized by many people who tell of seeing one of the creatures. The ears do not show. Arms and legs, although extremely wide and thick, do not show any bulging muscles, but there are heavy buttocks—something no ape has. The foot comes down on the ground heel-first, not flat or up on the toes.

Persons seeing the film, even "trained observers" with qualifications in zoology, almost always get the impression that the creature walks just like a man and has a build that would not be abnormal for a man. Frame-by-frame study of the movie and measurements of the proportions of the figure contradict both these impressions. The stride is actually much smoother than a normal man's, because the knee bends as the weight comes on it. A walking man bobs up and down as his body goes over the top of his straightened leg. The sasquatch in the film moves in a flowing fashion, with her legs being bent at all times. The legs are much straighter when she is reaching out in full stride than when one is bearing her full weight.

No precise measurements are possible, but taking as a standard the length of the foot, which is known from the tracks to be 14½ inches, the creature measures about seven

The sasquatch on the Patterson film.

feet in height and not much less than three feet across the shoulders. This is half again as wide as a heavily-built man, and other dimensions are proportionately heavy. Her thigh is as big as a normal man's chest, her ankle as big as his thigh. Her arms are long enough to span close to nine feet; two feet more than her height, but her body is also very long, so that her arms do not appear to hang very low. Her legs are shorter than those of a normal man. These dimensions cannot be taken as typical for all sasquatch, since both apes and man show extreme variations in body shape, but they do add to the problems for those who would like to dismiss the film as a hoax.

To begin with, the film itself has not been tampered with. Hollywood, as everyone knows, can produce virtually anything on film, including the same person playing two parts at the same time or imaginary monsters throwing around real people. These "special effects," however, are done in the film laboratories. Rene and I had some work done on Roger's film at one time, and the technicians at the film laboratory told us that it was definitely an original, not tampered with in any way. In other words, what the film shows is something that actually walked in front of the camera.

Such a thing could, in theory, be a machine, but I was told at the Disney studios in 1969 that it would have to be a machine more

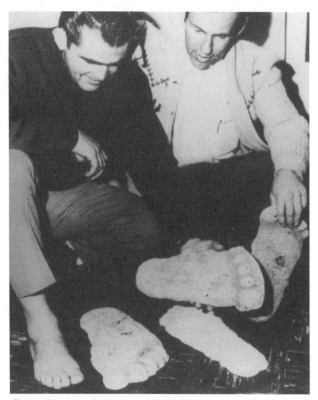

Bob Gimlin (left) and Roger Patterson, with casts of the tracks of the creature in the film.

74

ROGER PATTERSON'S MOVIE

sophisticated than any thing they could make. If it is not an animal or a machine then it would have to be a man in a suit. I showed the film to the man recommended by the Disney people as the best ape-imitator in Hollywood, and he said that it would have to be a skin-tight suit, not a padded one like his own gorilla suits, so there would have to be a man big enough to fill it. However it is not built like a man. It is far wider and deeper for its height.

I noticed that scientists who talked to Bob and Roger after seeing the film were not prepared to accuse them of perpetrating a hoax, but since they were also unable to accept that the animal could be real they seized on the only other alternative, that someone was out there in a fur suit hoaxing Bob and Roger. That's a nice easy explanation, except that it doesn't match the facts. Roger showed movies not only of the animal but also of its tracks, and those shots demonstrated that the sasquatch tracks sank in the ground to a far greater depth than the men's feet did as they moved around them. It is obvious from the depth that the big tracks were made by something several times heavier than the people. Now if there was a man swinging along in a suit when Roger took his movie it was most obviously not a man carrying a heavy load—not walking more than four miles an hour with 3½ foot strides and on deeply bent knees. A man carrying the weight to make the tracks that Bob and Roger said they saw the animal make would have to be a mighty individual to manage even a slow shuffle. It follows that if the thing on the film is a hoax the tracks and the film could not possibly have been made at the same time as Bob and Roger said they were, therefor Bob and Roger have to be involved in the hoax.

Before Roger died in 1972 he and Bob had a serious falling out, but at no time did . either one of them ever waver in their agreement that what they saw and photographed was genuine.

Also somewhat to the point is what Roger did after he got the movie. He came out to Willow Creek the same afternoon and through Al Hodgson immediately got word to Don Abbott at the British Columbia museum. Roger knew that I had brought a tracking dog with me the last two times I was in California and after mailing the film to his brother-in-law, Al deAtley, in Yakima, for private processing, he and Bob went back to Bluff Creek to await the arrival of the scientists and the dog.

I did my best to get some scientist to go down there, but none would make the trip.

The general reaction was that they would know all about it when they saw the film, if it showed anything. I spent so much time and money trying to round up scientists that I didn't have enough left of either to go to California myself, let alone hire a tracking dog. Meanwhile the weather broke in northern California and a deluge drove Bob and Roger out of Bluff Creek—not because they minded getting wet, but because they knew they were in danger of having their truck trapped behind slides. When they got out Rene had just arrived—he had been in San Francisco when he got the news —and Roger immediately asked him, "Where's Green with those scientists and dogs?"

Roger had no way of knowing that I would not be able to get there, or that there was going to be a cloudburst. If he were involved in a hoax he might have laid plans to fool the scientists, but it is hard to see how he could have been sure the dog wouldn't spoil things. With a track so fresh and undisturbed (Roger and Bob had not dared to follow the thing into the brush) a dog could hardly fail to get the scent, and if it wasn't able to lead the handler to some further tracks or some other evidence that it was following a creature like that on the film the whole business would quickly have started to smell more of fish than sasquatch.

A few days after getting the film processed Roger came to Vancouver to show it at the University of British Columbia, and Bob Titmus came down from Kitimat to see it. He went on from there to Bluff Creek to see what he could learn on the spot. Here are the comments he made in a letter to me·

My first full day up near the end of Bluff Creek, I missed the tracks completely. I walked some 14 to 16 miles on Bluff Creek and the many feeder creeks coming into it and found nothing of any particular interest other than the fact that Roger and Bob's horse tracks were everywhere I went. I found the place where the pictures had been taken and the tracks of Bigfoot the following morning. The tracks traversed a little more than 300 feet of a rather high sand, silt and gravel bar which had a light scattering of trees growing on it, no underbrush whatever but a considerable amount of drift debris here and there. The tracks then crossed Bluff Creek and an old logging road and continued up a steep mountainside....

This is heavily timbered with some underbrush and a deep carpet of ferns. About 80 or 90 feet above the creek and logging road there was very plain evidence where Bigfoot had sat down for some time among the ferns. He was apparently watching the two men

ROGER PATTERSON'S MOVIE

below and across the creek from him. The distance would have been approximately 125-150 yards. His position was shadowed and well screened from observation from below. His tracks continued on up the mountain but I did not follow them far. I also spent little time trying to backtrack Bigfoot from where his tracks appeared on the sandbar since it was soon obvious that he did not come up the creek but most probably came down the mountain, up the hard road a ways and then crossed the creek onto the sandbar. It was not difficult to find the exact spot where Roger was standing when he was taking the pictures and he was in an excellent position.

I spent hours that day examining the tracks, which, for the most part, were still in very good condition, considering that they were 9 or 10 days old. Roger and Bob had covered a few of them with slabs of bark etc., and these were in excellent condition. The tracks appeared perfectly natural and normal. The same as the many others that we have tracked and become so familiar with over the years, but of a slightly different size. Most of the tracks showed a great deal of foot movement, some showed a little, and a few indicated almost no movement whatever. I took plaster casts of ten consecutive imprints and the casts show a vast difference in each imprint, such as toe placement, toe gripping force, pressure ridges and breaks, weight shifts, weight distribution, depth, etc. Nothing whatever here indicated that these tracks could have been faked in some manner. In fact all of the evidence pointed in the opposite direction. And no amount of thinking and imagining on my part could conceive of a method by which these tracks could have been made fictitiously.

Most of us had assumed that the Bluff Creek road would be closed until the following spring, as that was what usually happened once the fall rains started slides.

As Bob discovered, the road was still

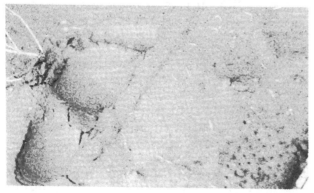

Lyle Laverty's photo of one of the tracks at the film site. Note shallow boot print at right.

open, and we later learned that a man named Lyle Laverty had taken pictures of some of the tracks before Bob got there. Jim Mc-Clarin, an active sasquatch hunter at that time and for several years afterwards, was a student at Humboldt State University at Arcata, about 80 miles from the film site, so he also was able to go and look it over.

I wasn't able to get there until the following June. By that time Rene and I had made a deal for Canadian lecture rights to the film, so I had slides of several frames from it with me. It was my intention to make a similar film with a man walking where the sasquatch had been. Jim, who is six foot five, and who had seen where the tracks went, was to be the sasquatch. I also got Roger's permission to show the film at Willow Creek, as the people there had never seen it, and on that occasion I met for the first time Ken Coon, who is still one of the most active investigators, and George Haas, whose *Bigfoot Bulletin* was the first and so far the best of the many small publications that have focussed on this subject.

The location proved to be just a few hundred yards upstream from the sandbar where Rene and I had seen tracks a few weeks before the film was taken. To our surprise a few of the depressions left when the tracks had been cast could still be seen, although none were evident in the critical area shown in the few clear frames of the movie. I found that making things in the background and foreground line up in the camera viewfinder as they appeared in the slides was a remarkably accurate way of locating the camera position. Moving the camera less than a foot in any direction would completely change the relationships. I presume that there may have been inaccuracies caused by differences between the lens of Roger's camera and my viewfinder, but I think I must have been very close to the right position.

It turned out that Roger had been close to the ground when he took the good footage, apparently squatting, since Bob found no marks to indicate that he had knelt. When he took the later shots of the animal from behind as it walked away he was still in exactly the same place, but standing up. He had never moved his feet after the sasquatch turned to look at him.

The film jiggles during the later sequence so we had assumed that Roger was again pursuing the creature, but apparently he was just trying to wind the camera and take movies at the same time.

While the camera was held close to the ground the slight undulations in the sandbar blocked any view of the exact points of contact between the creature's feet and the

ROGER PATTERSON'S MOVIE

These tracings from the two films, with background and foreground points lined up, give a close comparison of the size of Jim McClarin (6' 5" plus boots, 180 lbs.) and the sasquatch.

ground, so there is no way to be sure that Jim walked exactly where the sasquatch walked. There are also difficulties in lining up points in the background and foreground of the two movies because they were not taken at the same time of year or the same time of day, so the colors and the shadows are different. At one time I was certain that the creature was several inches taller than Jim, but now I am aware that it is possible to line up some frames so that Jim appears as tall or taller. It is still my opinion that the creature is shown to be the taller of the two, but the film does not establish that beyond question.

To the best of my knowledge no North American scientist or technician with any relevant expertise has ever studied the Patterson movie, but Rene succeeded in having it studied by two such men, one in England and one in the U.S.S.R., both specialists in biomechanics. The Englishman, Dr. D.W. Grieve, felt that if the film were taken at 24 frames or more per second the walk was similar to that of a human, but if it were taken at a slower camera speed then "a normal human being could not duplicate the observed pattern." Unfortunately the camera speed is not known.

The Russian, Dr. Dmitri Donskoy, found that the object filmed had a very efficient pattern of locomotion, quite different from any used by humans; that all its movements indicated much greater weight and strength than that of a human, and that there was no indication of an attempt to move in an unnatural way. The two men did not agree with each other, one being certain the thing was not human, the other unable to give a definite answer, but neither was able to reject it as an obvious fake, as so may less-qualified "experts" have found it so easy to do.

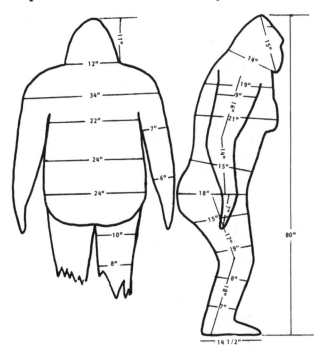

Approximate dimensions of the film sasquatch, based on the known length of the footprints.

Where Are The Bones?

It is often suggested that if tracks can be followed, then hair and droppings can be found and examined. This has in fact been done, but with inconclusive results. I was present when droppings of unusual appearance were collected in northern California and shipped off for examination. The report that came back was that the material was the remains of fresh water plants, and that it contained eggs of parasites otherwise known only from some North American tribal groups in the northwestern U.S., pigs from south China, and pigs and people from southwest China. Not very helpful, since the material obviously had not originated with either people or pigs. I have seen similar droppings in other places where they were thought to have been made by bears, and I am not convinced that such was not the case in this instance. However it is logical that the droppings of these creatures should be similar to those of bears eating similar diets, and unless the sasquatch conceals its droppings, which has been reported but is not normal primate behavior, such material must have been seen all along but misidentified as coming from some other animal.

Other ventures into the analysis of fecal matter proved equally inconclusive. The most recent that I was involved in concerned some droppings found by a game guide, Art German, at the scene of a sighting near Sexsmith, Alberta, in October, 1975. Bob Moody, of Woking, Alberta, a trapper with 45 years experience in the bush, said that as he was driving towards Sexsmith about 8:30 in the morning he saw beside the road a creature seven or eight feet tall and hair-covered, accompanied by two smaller ones only three or four feet high. Mr. German wrote to me that he had found toe marks dug into the bank by the road and he had also found three piles of fibrous droppings. At his request I had some of this material analyzed, but what the report boiled down to was that it contained grass.

The plain fact is that there is no need in this world for laboratory specialists who identify animals from their droppings, so such specialists don't exist.

The situation with hair is equally inconclusive, even though hair identification is a skill that does have qualified practitioners. The problem with hair is that it is identified by matching it with hair of known identity. If you can't find a matching sample, then you have an unidentified hair.

Some of the many hairs that have been collected at places where sasquatches were reported have been examined by qualified people and found not to match hair from any known North American mammal. That doesn't prove the hair came from a sasquatch, it only means that it wasn't proved that it didn't.

A guide named Wayne Twitchell in June 1967 found half a dozen hairs on a bush at a place on French Creek, in Idaho, where another man had reported seeing two silver grey, eight-foot humanoid creatures. The hairs were sent to Professor Ray Pinker, at California State College in Los Angeles, an instructor in police science with 36 years experience in a crime laboratory. Professor Pinker told me that there were two coarse guard hairs and half a dozen wool hairs, all of very fine structure, and that they did not match any of the hair samples available to him. He said that they resembled animal hairs in that they had naturally tapering tips and were darker in colour at the base than at the tip, whereas human hair is of uniform thickness and colour throughout.

On the other hand, most strands had no medulla in the center, which is characteristic of human hair, but also of some sheep and goats. Most animal hairs have a pronounced medulla. Also the scales on the outside of the hairs very much resembled characteristic scales of human hairs.

That sort of report is interesting, even encouraging, but obviously falls far short of providing a way to prove the existence of an unknown animal.

There is one glaring omission from the list of physical evidence, namely bones. It is natural to assume that even if a type of animal had managed to avoid being killed or captured it could not be present in sufficient numbers to survive without its bones being found from time to time.

There is one possible answer to that problem—that the sasquatch have reached such a state of intelligence that they deliberately conceal their dead. There are even some Indian stories that suggest that. For my own part I doubt it very much, as I feel that it is inconsistent with the general pattern of behavior exhibited by these animals.

Another possibility is that they seek hiding places before they die. That does not seem improbable, but there would be exceptions in case of violent deaths, and so accidental discoveries of bones would still be expected.

The anomaly is not as great as it seems, because the liklihood of finding bones is not as great as most people think. Anyone who has spent time in the woods will realize that bodies and bones of animals that have died natural deaths are very seldom seen,

WHERE ARE THE BONES?

and that even the finding of a single piece of bone is not an everyday occurence. Considering the large number of animals that must die each year it is obvious that an efficient bone-disposal system must exist in nature. In fact bones form a part of the diet of many living things both large and small, and they seldom last very long.

There are many reports of "giant burials" from various places in North America, and the fact that there is no generally-known record of any bones from these reposing in any of the thousands of museums on the continent does not establish that they are not in fact to be found there. Material for which no one currently on the staff of a museum has any use tends to get put away and forgotten. At one time I was of the opinion that some museum somewhere probably had some sasquatch bones, but I no longer consider it likely. A creature built like the one in the movie, and as heavy as is indicated by the footprints, would have bones far to massive to be mistaken for human bones, however similar they might be in other ways.

There is, however, a lead to some bones that someone might be able to find. In June, 1971, the Salem, Oregon, *Capital Journal* carried a story about a local girl and her friend having seen a dead Bigfoot near Happy Camp, California, in 1967. I didn't learn of the story until several years later, and didn't pay much attention to it. For one thing, it contained the statement, "We had walked about eight miles into the woods," and the idea of two girls being able to find any particular spot that far off in the brush the same day, let alone years afterwards seemed most unlikely. Still, I never felt quite right about leaving such a story uninvestigated, and I tried a few times to locate the girl or to get someone else to do it. Late in 1976 I finally took the matter up in earnest, and got lucky. Within 24 hours I had met one of the girls, now married and living near Tacoma, Washington, and had talked to the other on the phone in Norco, California.

It turned out that they had not been wandering in the woods at all, but had been walking on an old forest road. In other words, whether they could find the exact spot or not, they could at least walk right over it.

Their stories differed considerably. One of them was quite sure she had seen a Bigfoot, lying on its back with its legs spread out. She recalled trying to turn it over by prying at it with a stick and finding that it was too heavy. The other girl could

remember only a great deal of hair, like a horse's mane except that there was too much of it and it was spread over too large an area, with the body all rotted away. She decided she must have seen some bones too, because it was from the bones that she got the impression that it was a big animal. She did not know of any animal it could have been.

The first girl could not remember any face, she thought that and the stomach area was rotted away, and she remembered weathered white ribs showing. Ignoring the great difference in their general impression and concentrating on the details described there were no actual contradictions.

The two of them had been neighbours for a time at Salem, then one had moved to Kemper Gulch, 10 miles north of Happy Camp in the Bigfoot country of northern California, and the other had later come to visit her there. One day they took a walk several miles up the road leading to the old Wagner Ranch. The girl I talked to first was sure she had been there in August, yet insisted that the animal was lying in several inches of old snow. The second girl, who lived there, said her friend visited twice, and this happened in the Easter holidays. She thought the remains were a bit below the road. The first girl thought they were on it.

I am entirely satisfied that there was indeed what remained of some big-boned animal on or by that road, and I am pretty sure that at least the second girl, who knew the area, should be able to take someone pretty close to the spot. Even so many years afterward I think there is an excellent possibility that some bone or tooth from an animal that big will have survived, although it might take many hours to find any. As to what the bones would be from, there is one promising indication. The first girl said that the thing that made the most impression on her was seeing a hand with a thumbnail, a nail larger than a human one, and thick, but squared off; definitely a nail, not a claw or hoof. Anything with a nail would have to be a primate.

There have been several attempts to find the bones the girls described, but none at time of writing have involved the time and manpower that would be necessary to have any real hope of success or to establish that no such bones are there.

And even if the sasquatch has none of his modern bones available for examination he is at least ahead of his cousin the gorilla in one respect. Remains of a suitable ancestor for the gorilla have never been found, but for the sasquatch there is one, and possibly two. There are a number of people who

WHERE ARE THE BONES?

speculate that *Australopithecus robustus,* alias *Paranthropus,* might have grown larger during the millions of years since he left his bones in Africa. His presumed lifestyle would be suitable, and he had a crest on his skull that would provide the peaked head of the Patterson sasquatch. The more probable candidate, in my opinion, is one that would not require any change in size, but otherwise not much is known about him. He is represented only by a number of loose teeth and four lower jaws, all of such awesome proportions that he was given the name *Gigantopithecus.* The teeth and jaws obviously belong to an ape, but are about twice the size of those of a gorilla. Whether the body was in proportion to the jaw, or whether the animal went upright or on all fours, will not be known until some other bones turn up.

Gigantopithecus fossils have been found in India and China, so speculation that he eventually reached the land bridge to North America is not unreasonable. There is no proof that he has any modern descendants, or that the sasquatch, even assuming it to be a real animal, is such a descendant. But it is established that there have been apes far bigger, at least in the jaw, than any known to exist today.

Does it seem impossible that anything so large could be so elusive? In 1872 the last dozen or so of the Yahi Indians gave up fighting the white man and disappeared in a small valley on the slopes of Mount Lassen in California. They lived there undetected for 12 years, until they started raiding cabins for food, and in a decade of cabin-raiding they were seen only once.

In 1894 the last four survivors made their home on two pieces of land no more than a half mile wide and three miles long. Their presence was unknown for 15 years until some surveyors blundered into their camp. When the last survivor walked into captivity in 1911 he had lived for 39 years in concealment on the fringe of civilization, yet his people had permanent camps, used fires, and even cremated their dead.

By contrast the sasquatch, with no homes or fires and unlimited space to roam in, are frequently seen. It isn't so much that they elude us as that we delude ourselves.

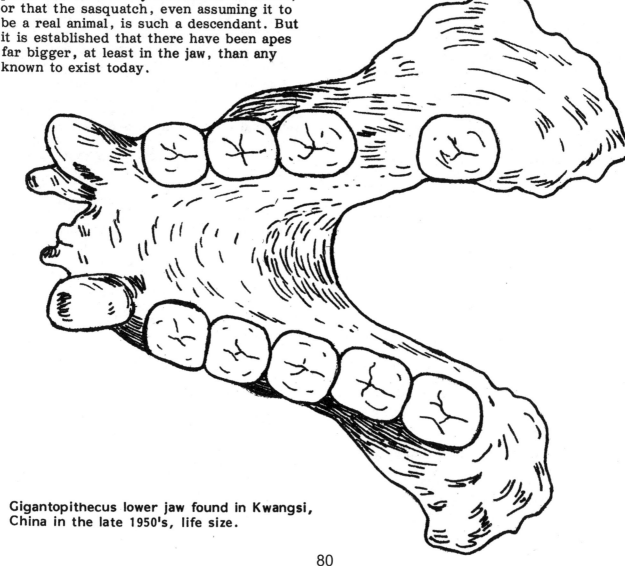

Gigantopithecus lower jaw found in Kwangsi, China in the late 1950's, life size.

80

Giants from the Past

In October, 1967, Roger Patterson and Bob Gimlin emerged from the Bluff Creek valley in the northwest corner of California with a short strip of 16 mm. movie film that showed an upright creature almost seven feet tall, but looking like a huge ape, striding across a sandbar. Everything about the sasquatch investigation has been different since that time.

The movie became by far the most significant single piece of evidence that there are giant bipedal apes living in western North America, and although science has chosen to ignore it no one has ever been able to establish that there is anything the matter with it.

Before the movie hit the headlines only a minority of people were aware that there were reports of hairy giants being seen in the vast forested areas of western North America, and that giant tracks resembling human footprints had been found, photographed and cast from time to time.

There had been a flare-up of interest in 1924 when a group of miners claimed to have been bombarded by giant hairy apes on Mount St. Helens in southwestern Washington, and there was some degree of awareness in British Columbia of Indian traditions of hairy "Sasquatch" giants that occasionally came down from their mountain retreat.

In 1958 there was another burst of publicity when 16-inch "Bigfoot" tracks first started to appear at Bluff Creek. From then onward a small number of people continued to be active in investigating both the footprints and an increasing number of modern reports of encounters with hairy giants, but their efforts did not attract much public attention.

The public interest stirred up by the Patterson movie resulted in an information explosion. I have no way of knowing how many people wrote to Roger or contacted him in other ways to tell of their own experiences with sasquatches or what they thought were sasquatches. Certainly there were hundreds. At the same time a considerable number of people joined the ranks of the active investigators, and although there was generally more rivalry than co-operation a lot of information was passed around. People who had been keeping quiet for years for fear of ridicule, or in some cases had convinced themselves that they couldn't really have seen what they saw, were now ready and even eager to tell somebody. So were people who had just seen something the day before,

and in most areas there was some enthusiast who had been given enough publicity so that these people could make contact.

The new situation was not altogether an improvement, because along with the stories that were apparently genuine came a fair number that were plainly wishful thinking, and some that were deliberate lies. A particular problem was created with regard to track reports. In the early days we would not hesitate to travel hundreds of miles to check out a track report because we could be almost certain of seeing something interesting. With widespread publicity that situation reversed itself completely. So many people began getting excited over miscellaneous holes in the ground that track reports became a very bad gamble, not worth committing any time and money to. Asking for detailed descriptions was no help, people could see toe marks in a cow track if toes were what they wanted to see. Once I drove 700 miles to inspect a series of sasquatch tracks of different sizes and ages indicating that a whole group had been going down to the river at the same place regularly for many days—and when I got there I couldn't see a single thing that remotely resembled a track. And one sure result of that sort of experience is that there must have been times when I decided not to go when there was really something to see. Another result, for a person like myself who keeps a complete file of reports, is the knowledge that the reliability of those files is a lot lower than it used to be. A lot of reports that were never investigated shouldn't be there at all, but which ones?

Under those circumstances I developed a fondness for the stories that people were finding in newspaper files and other old records. When someone reports a sighting there is no way of being sure what, if anything, they saw. Tracks are better, if you get there in time, but unless they are kept secret they don't last long. The people who flock to see them inevitably step on them. Printed reports from any time before the sasquatch publicity started cannot be the result of that publicity, and they don't go away. The earliest stories that have turned up are from places in the eastern United States—they date from before the West had any newspapers—but over the years a file has accumulated for each of the Pacific Coast states as well as for British Columbia.

The very early western reports are all references to Indian traditions, and while those are well worth a book in themselves they are not the subject of this one. An example that is worth noting, considering what has happened since, is to be found in

GIANTS FROM THE PAST

the book *Wanderings of an Artist*, by Paul Kane. He quotes from an entry in his journal for March 26, 1847:

When we arrived at the mouth of the Kattlepoutal River twenty-six miles from Fort Vancouver, I stopped to make a sketch of the volcano, Mount St. Helen's, distant, I suppose about thirty or forty miles. This mountain has never been visited by either whites or Indians, the latter assert that it is inhabited by a race of beings of a different species, who are cannibals, and whom they hold in great dread. . . . These superstitions are taken from the statement of a man who, they say, went to the mountain with another and escaped the fate of his companion, who was eaten by the "Skookums" or evil genii. I offered a considerable bribe to any other Indian who would accompany me in its exploration, but could not find one hardy enough to venture.

The sketch to which the artist refers shows the mountain in the process of erupting, not from a crater in the peak but from one side, and indeed St. Helen's was an active volcano at the time when Paul Kane saw it. Today everyone knows it as an active volcano again, but for more than half a century its chief claim for recognition was not its fiery past, which was assumed to be past forever, but its 'apes.'

The earliest reference to an encounter by a settler with what we would now call a sasquatch also refers to Mount St. Helen's. In *Told by the Pioneers*, a volume of interviews with old timers in Washington State during the 1930's, Agnes Louise (Ducheney) Eliot is quoted as follows:

Grandpa Ducheney firmly believed the story of the huge apes near St. Helens Mountain. He went there to hunt once and one of these apemen beckoned to him. He just turned and ran until he reached home.

There is no date attached to that story, but the context indicates that "Grandpa Ducheney" was actually Mrs. Eliot's father, and that he died while his children were still young. She also notes that General Grant had stayed with the Ducheneys, which would have to have been between 1851 and 1853. The other candidate for the oldest story from the Northwest would be one told by another president, Theodore Roosevelt, which will be quoted in full in a later chapter.

The earliest stories with a definite date that have so far come to my attention were from the vicinity of San Francisco, in 1870. The series began with a small article in the San Joaquin *Republican* on September 19, in a column of "Local Brevities," as follows:

WILDMAN—We learn from good authority that a wild man has been seen at Crow Canyon, near Mount Diablo. Several attempts have been made to capture him, but as yet have proved unsuccessful. His tracks measure thirteen inches.

That little item stirred the Oakland *Daily Transcript* to respond on September 27:

An item appeared in the San Joaquin Republican the other day stating that a wild man had been seen in some part of the San Joaquin county, and we afterward noticed the statement copied into several other papers, with brief comments indicating disbelief in the report. We must confess to a want of credulity on our part also as to the exact correctness of the item at the time, but we were yesterday placed in possession of certain information which leads us to believe that there may be some foundation for the report. As our columns are somewhat crowded this morning, we will give the reports as we received them as briefly as possible.

F. J. Hidreth and Samuel De Groot, of Washington Corners, in this county, while out hunting on Orias Timbers Creek in Stanislaus county about three weeks ago, discovered footprints along the bank of the creek resembling the impressions of a human being's feet. Mr. Hildreth, who gave us this information, states that the tracks were like those of a human being with the exception that the impressions of the toes were much larger.

Hildreth afterward became separated from his companion, and upon proceeding some distance up the creek, saw a few yards ahead of him what he believed to be gorillas. If the description Mr. Hildreth has given us of these animals is true, he is certainly warranted in believing them to be of that species of animal. Mr. De Groot also reports that he saw the same objects and is positive that they are gorillas. The appearance of these strange animals in that neighborhood is notorious and that they are gorillas is firmly believed by a great many people in that vicinity.

A number of old hunters have started out to capture them, and we are promised whatever further facts may occur as soon as the party returns. The above we gathered from various parties, and whether true or not, there are many persons in the neighborhood of Washington Corners who firmly believe that the animals referred to are veritable gorillas.

This is a frame from the Patterson movie, the only known photograph of a sasquatch.

This, for comparison, is what a bear looks like standing on its hind legs.

That story was reprinted on October 1 in the Antioch *Ledger,* and as a result the *Ledger* received the following letter from a correspondent in Grayson, California:

I saw in your paper a short time since an item concerning the 'gorilla' which was said have been seen in Crow Canyon and shortly after in the mountains at Orestimba Creek. You sneered at the idea of there being any such a 'critter' in these hills, and, were I not better informed, I should sneer too, or else conclude that one of your recent prospecting party had got lost in the wilderness, and did not have enough sense to find his way back to Terry's.

I positively assure you that this gorilla or wildman, or whatever you choose to call it is no myth. I know that it exists, and that there are at least two of them, having seen them both at once not a year ago. Their existence has been reported at times for the past twenty years, and I have heard it said that in early days an ourang-outang escaped from a ship on the southern coast; but the creature I have seen is not that animal, and if it is, where did he get his mate? Import her as the web-foot did their wives?

Last fall I was hunting in the mountains about 20 miles south of here, and camped five or six days in one place, as I have done every season for the past fifteen years. Several times I returned to camp, after a hunt, and saw that the ashes and charred sticks from the fireplace had been scattered about. An old hunter notices such things, and very

soon gets curious to know the cause. Although my bedding and traps and little stores were not disturbed, as I could see, I was anxious to learn who or what it was that so regularly visited my camp, for clearly the half burnt sticks and cinders could not scatter themselves about.

I saw no tracks near the camp, as the hard ground covered with leaves would show none. So I started in a circle around the place, and three hundred yards off, in damp sand, I struck the track of a man's foot, as I supposed—bare and of immense size. Now I was curious, sure, and I resolved to lay for the barefooted visitor. I accordingly took a position on a hillside, about sixty or seventy feet from the fire, and securely hid in the brush, I waited and watched. Two hours and more I sat there and wondered if the owner of the feet would come again, and whether he imagined what an interest he had created in my enquiring mind, and finally what possessed him to be prowling about there with no shoes on.

The fireplace was on my right, and the spot where I saw the track was on my left, hid by the bushes. It was in this direction that my attention was mostly directed, thinking the visitor would appear there, and besides, it was easier to sit and face that way. Suddenly I was surprised by a shrill whistle, such as boys produce with two fingers under their tongues, and turning quickly, I ejaculated, "Good God!" as I saw the object of my solicitude standing beside my fire, erect, and looking suspiciously

GIANTS FROM THE PAST

around. It was the image of a man, but it could not have been human.

I was never so benumbed with astonishment before. The creature, whatever it was, stood fully five feet high, and disproportionately broad and square at the fore shoulders, with arms of great length. The legs were very short and the body long. The head was small compared to the rest of the creature, and appeared to be set upon his shoulders without a neck. The whole was covered with dark brown and cinnamon colored hair, quite long on some parts, that on the head standing in a shock and growing close down to the eyes, like a Digger Indian's.

As I looked he threw his head back and whistled again, and then stooped and grabbed a stick from the fire. This he swung round and round, until the fire on the end had gone out, when he repeated the manoeuver. I was dumb, almost, and could only look. Fifteen minutes I sat and watched him as he whistled and scattered my fire about. I could easily have put a bullet through his head, but why should I kill him? Having amused himself, apparently, as he desired, with my fire, he started to go, and, having gone a short distance he returned, and was joined by another—a female, unmistakably— when both turned and walked past me, with-

in twenty yards of where I sat, and disappeared in the brush.

I could not have had a better opportunity for observing them, as they were unconscious of my presence. Their only object in visiting my camp seemed to be to amuse themselves with swinging lighted sticks around. I have told this story many times since then, and it has often raised an incredulous smile; but I have met one person who has seen the mysterious creatures, and a dozen of whom have come across their tracks at various places between here and Pacheco Pass.

Mount Diablo is just east of Oakland. Orestimba Creek is about 50 miles to the southeast. Twenty miles south of Grayson indicates the same vicinity as Orestimba Creek.

There are no stories from the time of the gold rushes, either in California in '49 or British Columbia in '58, but a reference to that period is to be found in a letter written to *True* by John M. Weeks of Providence, R.I. after the magazine printed Ivan Sanderson's article about the California Bigfoot in December, 1959:

My grandfather prospected for gold in the eighteen fifties throughout the region described as being the home of the Snowman. Upon grandfather's return to the East he

In Bigfoot country two feet make a yard. Bruce Berryman, Bob Titmus and Syl McCoy show casts of some of the big footprints found near Hyampom, California, in April, 1963. Tracks like these are a major element in the sasquatch mystery. Something has to make them.

GIANTS FROM THE PAST

told stories of seeing hairy giants in the vicinity of Mount Shasta. These monsters had long arms but short legs. One of them picked up a 20-foot section of a sluiceway and smashed it to bits against a tree.

When grandfather told us these stories we didn't believe him at all. Now, after reading your article, it turns out he wasn't as big a liar as we youngsters thought he was.

In general this book does not deal with southern California, but the second oldest newspaper account from the Pacific coast comes from about as far south as you can get. It was printed in the San Diego *Union* on March 9, 1876, under the heading:

A WILD MAN IN THE MOUNTAINS

The following strange story is sent us by a correspondent at Warner's Ranch in this county. We know the writer to be a perfectly reliable person and believe his statement, singular though it may seem, to be fully entitled to credence:

WARNER'S RANCH—March 5—About ten days ago Mr. Turner Helm and myself were in the mountains about ten miles east of Warner's Ranch, on a prospecting tour, looking for the extension of a quartz lode which had been found by some parties sometime before. When we were separated, about half a mile apart—the wind blowing very hard at the time—Mr. Helm, who was walking along looking down at the ground, suddenly heard someone whistle.

Looking up he saw 'something' sitting on a large boulder, about fifteen or twenty paces from him. He supposed it to be some kind of an animal, and immediately came down on it with his needle gun. The object instantly rose to its feet and proved to be a man. This man appeared to be covered all over with coarse black hair, seemingly two or three inches long, like the hair of a bear; his beard and the hair of his head were long and thick; he was a man of about medium size, and rather fine features—not at all like those of an Indian, but more like an American or Spaniard.

They stood gazing at each other for a few moments, when Mr. Helm spoke to the singular creature, first in English and then Spanish and then Indian, but the man remained silent. He then advanced towards Mr. Helm, who not knowing what his intentions might be, again came down on him with the gun to keep him at a distance. The man at once stopped, as though he knew there was danger.

Mr. Helm called to me, but the wind was blowing so hard that I did not hear him. The wild man then turned and went over the hill and was soon out of sight; before Mr. Helm could come to me he had made good his escape. We had frequently seen this man's tracks in that part of the mountains, but had supposed them to be the tracks of an Indian. I did not see this strange inhabitant of the mountains myself; but Mr. Helm is known to be a man of unquestioned veracity, and I have no doubt of the entire truth of his statement.

L.T.H.

The earliest Oregon report I have was sent to me by Tim Church, a researcher from Rapid City, South Dakota. It comes from the Carson City, Nevada, *Morning Appeal*, December 31, 1885, and was headed, not very originally:

WILD MAN IN THE MOUNTAINS

Much excitement has been created in the neighborhood of Lebanon, Oregon, recently over the discovery of a wild man in the mountains above that place, who is supposed to be the long lost John Mackentire. About four years ago Mackentire, of Lebanon, while out hunting in the mountains east of Albany with another man, mysteriously disappeared and no definite trace of him has ever yet been found. A few days ago a Mr. Fitzgerald and others, while hunting in the vicinity of the butte known as Bald Peter, situated in the Cascades, several miles above any settlement saw a man resembling the long-lost man, entirely destitute of clothing, who had grown as hairy as an animal, and was a complete wild man.

He was eating the raw flesh of a deer when first seen, and they approached within a few yards before he saw them and fled. Isaac Banty saw this man in the same locality about two years ago. It is believed by many that the unfortunate man who was lost became deranged and has managed to find means of subsistence while wandering about in the mountains, probably finding shelter in some cave. A party of men is being organized to go in search of the man.

There are two more Oregon reports from before the turn of the century, but I do not have either one from anything approaching an original source. Ivan Sanderson, in *Abominable Showmen; Legend Come to Life*, tells of enormous human footprints being seen around a mining camp 50 miles inland in the Chetko River area and two men standing guard being smashed to death in the night. No source for the story is given, nor is there anything specifically pinning the violence on the sasquatch. I have a copy of another story, written by Dale Vincent in 1947, which tells of two miners prospecting above the Sixes River in 1899 seeing their camp gear thrown off a cliff by a big, powerful erect creature, "neither man nor beast" with yellow fuzz all over its body. They shot at it but it fled. Again no source is given for the information, and the clipping does not even show what paper it was from. Both these locations are in the southwest corner of Oregon, and there are two more specific reports from the Sixes River within the following five years. Both were originally published in the Myrtle Point *Enterprise*, then were copied in other papers. Here is the first as it appeared in the Roseburg, Oregon, *Daily Review*, on December 24, 1900:

A KANGAROO MAN—The Sixes mining district in Curry county has for the past 30 years gloried in the exclusive possession of a "kangaroo man." Recently while Wm. Page and Johnnie McCullock, who are mining there, went out hunting McCulloch saw the strange animal-man come down to a stream to drink. In calling Page's attention to the strange being it became frightened, and with cat-like agility, which has always been a leading characteristic, with a few bounds was out of sight.

The appearance of this animal is almost enough to terrorize the rugged mountain sides themselves. He is described as having the appearance of a man—a very good looking man—is nine feet in height with low forehead, hair hanging down near his eyes, and his body covered with a prolific growth of hair which nature has provided for his protection. Its hands reach almost to the ground and when its tracks were measured its feet were found to be 18 inches in length with five well-formed toes. Whether this is a devil, some strange animal or a wild man is what Messrs. Page and McCulloch would like to know, says the Myrtle Point Enterprise.

The second story was reprinted in the *Lane County Leader*, Cottage Grove, Oregon, on April 7, 1904:

SIXES WILD MAN AGAIN

Visits the Cabins of Miners And Frightens the Prospectors

At repeated intervals during the past ten years thrilling stories have come from the rugged Sixes mining district in Coos County, Oregon, near Myrtle Point, regarding a wild man or a queer and terrible monster which walks erect and which has been seen by scores of miners and prospectors. The latest freaks of the wild man is related as follows in the last issue of the Myrtle Point Enterprise:

The appearance again of the "Wild Man" of the Sixes has thrown some of the miners into

a state of excitement and fear. A report says the wild man has been seen three times since the 10th of last month. The first appearance occurred on "Thompson Flat." Wm. Ward and a young man by the name of Burlison were sitting by the fire of their cabin one night when they heard something walking around the cabin which resembled a man walking and when it came to the corner of the cabin it took hold of the corner and gave the building a vigorous shake and kept up a frightful noise all the time—the same that has so many times warned the venturesome miners of the approach of the hairy man and caused them to flee in abject fear.

Mr. Ward walked to the cabin door and could see the monster plainly as it walked away, and took a shot at it with his rifle, but the bullet went wild of its mark. The last appearance of the animal was at the Harrison cabin only a few days ago. Mr. Ward was at the Harrison cabin this time and again figures in the excitement. About five o'clock in the morning the wild man gave the door of the cabin a vigorous shaking which aroused Ward and one of the Harrison boys who took their guns and started in to do the intruder. Ward fired at the man and he answered by sending a four pound rock at Ward's head but his aim was a little too high. He then disappeared in the brush.

Many of the miners avow that the "wild man" is a reality. They have seen him and know whereof they speak. They say he is something after the fashion of a gorilla and unlike anything else that has ever been known; and not only that but he can throw rocks with wonderful force and accuracy. He is about seven feet high, has broad hands and feet and his body is covered by a prolific growth of hair. In short, he looks like the very devil.

At time of writing I am not aware of any such old newspaper reports from anywhere in the state of Washington. The earliest detailed account I have concerns an incident that took place in 1912, an account of which was contained in a letter written to me in 1969 by a lady who grew up on a ranch near Oakville in Gray's Harbor County.

After giving some of her family history and mentioning that an old Indian chief, Pike Ben, had told her of seeing a sasquatch on an island between the Chehalis and Black Rivers, she continued as follows:

The spring of 1933 was a wet one and also I remember we had a lot of snow. All the ranches in our area were on the river. We all farmed the bottom land but built our houses on the hillside. This area had been logged by the Balch Logging Company in

GIANTS FROM THE PAST

the period of 1915–1923. By 1933 a lot of
the hill area had second growth timber on
it. I was attending High School in Oakville
at this time and this particular weekend I
had been to a dance in Oakville with a group
of other young girls. When I returned home
I always walked around the house to the
back door. All my life I had heard coyotes
and cougars. Just as I started to open the
back door I heard a very loud noise up on
the hill. I had never heard a noise like it
before and I was a little scared. I went in
and woke up my mother and told her to come
and listen as this was a noise I had never
heard before. Also I could hear another one
answer way over on another hill.

Mother got up and her reaction was so
strange that I could hardly believe it was
my mother. She kept saying, "It's that terrible Ape again." When I kept trying to ask
her to explain she kept saying "I'll tell you
tomorrow, and let's get inside as we are not
safe out here." This is the story my mother
told me the next day.

After my Grandfather died my Grandmother divided up the old ranch and gave
the children a Stump Ranch. About 40 acres
and most of my mother's land was the area

Two 16-inch prints at Hyampom, showing the
stride and the great variation in toe position

where our house stood. The timber came
down almost to the back of the house. She
said in the fall of 1912 my father had gone to
Aberdeen to sell potatoes he had raised on
the 10 acres they had cleared. He had to
stay all night so Mother was alone in the
small house.

The house had a porch across the front
and a large bay window. She said she had
not slept very soundly because she was not
used to being alone. She said about 1:00
a.m. she heard a terrible stomping noise on
the front porch. She got up to see what it
was. It was moonlight outside, and at first
she thought it was a bear on the porch, but
this animal was standing on its back legs and
was so large it was bending over to look in
the window. She said it appeared over 6 feet
tall and it didn't look like a bear at all in the
moonlight.

She said in a few minutes it walked over

GIANTS FROM THE PAST

and jumped off the front porch and started around the house. She went into the kitchen so she could get a good look and she said it looked just like an ape. She said the strange thing that had scared her most was the noise it made as it walked around the house.

She said as soon as it was daylight she went over to my Grandmother's place and told them what she had seen. Her brothers all made fun of her and told her she had had a nightmare. When my Father returned home he wouldn't believe her either. She said she would never mention it again. This was the first time I heard her tell the story.

The lady also mentions that when she and her husband had a summer home in the Bitter Root Mountains in Montana they occasionally heard old miners tell of apelike creatures in the area.

The story of the capture of "Jacko" at Yale in 1884 is the only one I know of from the British Columbia papers before the turn of the century, but on Vancouver Island starting in December, 1904, there was a series of reports that lasted several months and that focussed attention on an earlier report as well. The earliest story I have found was in the Victoria *Colonist*, December 15, 1904, under the heading:

THE WILD MAN GIVES AUDIENCE

The Vancouver Island wild man, whose existence has been persistently prominent in the Indian traditions of the northern island tribes, and who has been vouched for by Mike King, perhaps the most famous backwoodsman and timber cruiser of British Columbia, has again made his appearance, and this time there are four creditable witnesses to attest that he is no myth or phantom of Indian imagination.

A.B. Crump, J. Kincaid, T. Hutchins and W. Buss, four sober-minded settlers of Qualicum, are the new witnesses and there is not the slightest deviation or variation in detail in the stories they tell with an earnestness that defies ridicule. They were out hunting in the vicinity of Horne lake, which lies almost midway between Great Central lake and Comox lake, in an uninhabited section of the interior of Vancouver Island, when they came upon the uncouth being whom they describe as a living, breathing and intensely interesting modern Mowgli.

The wild man was apparently young, with long and matted hair and beard, and covered with a profusion of hair all over the body. He ran like a deer through the seemingly impenetrable tangle of undergrowth, and pursuit was utterly impossible.

It was three or four years ago that Mike King came across the strange creature, which, according to his description, comes as near to the missing link as anyone could imagine. Mr. King was at the time making one of his solitary tramps, prospecting the timber country inland from the head of Campbell river, his Indians having zealously refused to go beyond a certain landmark which they referred to as the boundary of the man-monkey's country.

It was drawing toward evening, and the man beast was surprised in the act of bending over a little waterhole washing certain edible grass roots, which he had disposed in two neat little piles—the one cleaned, the other awaiting cleaning.

At the sound of Mr. King's approach he uttered a very human cry of mingled terror and defiance, ran up the hillside half way and then stood to curiously regard the lonely intruder, who kept him covered with his rifle.

Mr. King's description was very similar to that given now by the Qualicum quartette and their meeting place with the wild man is virtually the same territory.

As soon as he had got his nerves together again, Mr. King made an effort to join the remarkable creature, which at his approach dashed through the underbrush exactly as a terrified deer would. He did not see him again, although his extraordinary cries were heard at intervals during the night, and Mr. King, who sat the night out by a big fire, rifle in hand, was thoroughly convinced that the wild man slipped back silently to inspect him at distant range during his vigil.

He says that the body was covered with reddish-brown hair and that the arms were peculiarly long and used freely in climbing and bush running while the trail showed a distinct human foot, but with phenomenally long and spreading toes.

The Indians all through the North Vancouver island shun what they have set apart as the wild man's country, and no money or persuasive art will induce them to go into it with hunter, trapper, timber cruiser or prospector.

Their story, which Mr. King wormed from an old native with much difficulty, is to the effect that at the coming of the Spaniards to the West coast an immense monkey (presumably an ape or ourang-outang) escaped from one of the vessels and took refuge in the forest wilderness: several months later it appeared suddenly at a West Coast village, and the Indians running in terror, it caught and carried away with it a girl of fifteen. The natives allege that two wild children resulted, but that one was found dead in a

GIANTS FROM THE PAST

hut beyond the headwaters of Campbell river years ago, leaving the other lonely creature to dominate the woods. Mr. King himself saw in the unexplored country shunned by all Indian tribes a hut that seemingly had been built for shelter by some semi-human denizens of the woods.

The Qualicum men say that several times during the past few years reports have come to the settlements of such a creature being seen, but as the truth of the encounter depended upon one or at most two witnesses they have been universally scouted. Mike King is a man whose word is absolutely good throughout British Columbia. He knows the wilderness thoroughly, is not at all imaginative, and is a strict teetotaller. On the sub ject of the wild man he is ready at any time to make sworn attestation to the exact truth of his report, but he has been reluctant to discuss the matter in view of the ever-ready joke.

There is a new theory evolved at Qualicum, which connects the wild man with remembrance of the fact that nine years ago a child was lost from the settlement, disappearing in the forest and never again emerging. They incline to the opinion that this child has lived, succoured by the beasts in Mowgli fashion, and is the wild man of today.

There is no doubt whatever of the historical authenticity of Mike King. The provincial library has an index of newspaper reports about developments in his timber and mining interests. On December 30, 1905, the Vancouver Province published an interview in which he recalled other details of what he had seen:

I came up by accident and was only about twenty-five feet away when it caught sight of me. The Mowgli, or whatever you like to call it, was squatted down like a monkey beside a little stream. It was washing a kind of wild onion that we prospectors sometimes eat when we run short of provisions. I thought it was a bear at first, but when it heard me and straightened up I brought my rifle up to my shoulder, for it was the strangest sight I ever expect to set eyes on.

Imagine meeting in the depths of a lonely forest an animal in the shape of a man, standing over six feet high, and covered all over with long, coarse black hair, which on some portions of the body was a foot long. The forehead was low and retreating, and its small eyes glared at me in surprise for an instant. Just below the eyes the hair on the face of the creature was short, but everywhere else it was long and shiny

and on the hand the hair hung down below the finger ends. The being stood quite straight for a moment in surprise, and seemed to me to weigh about two hundred and forty pounds. If it had taken one step toward me I would have sent a bullet through it, for I had it covered with my rifle. But after glaring at me for a moment it uttered a cry—a half-human sort of grunt—and grasping a branch near by, hoisted itself up the bank of the creek and ran away through the underbrush, slightly bending as it did so, with the speed of a startled deer.

I examined the creature's footprints afterward. Its feet were short and very broad. I noticed the heel came back almost to the point, like that of a gorilla. The armful of wild onions that it had been cleaning at the little stream was as nicely done up as if a human being had prepared them.

The next report of the wild man appeared in the Colonist on May 2, 1905, headed:

THE WILD MAN WOUNDED

When the tale of the wild man of Vancouver Island, the prototype of Kipling's Mowgli, was bruited abroad months ago, many residents of Qualicum were on record to vouch for their belief in the existence of this strange creature. Capt. Owens, the Nanaimo pilot, who has returned from Union, brings news of the wild man having again been seen, shot at, and wounded by Indians.

Mr. John Fraser, of the firm of Fraser & Howe, Union, told the pilot of the adventure of the Indians. They were in a canoe between Comox and Union Bay and near the beach when they saw what they believed to be a bear on the shore. One of the Indians raised his shotgun and fired. The object they believed to be a bear then straightened up and the Indians saw that it was a man, naked and covered with hair. He had been digging clams when they fired.

The shot evidently took effect for the wild man sprang up, yelled, and ran into the woods. The Indians did not follow. They were terrified and paddled quickly away.

They told of what they had seen on arriving at the home village and all the Indians of that section are greatly excited as a result. The natives who shot at the wild man are emphatic in their statements that his actions were similar to those of the wild man previously seen near Qualicum, for whom search parties have looked in vain.

Capt. Owens says the wild man is believed to be a young man who disappeared

GIANTS FROM THE PAST

from Qualicum twelve years ago. Capt. Owens was then master of the steamer City of Nanaimo and the young man, who stayed at the residence of Mr. Buss, went into the woods one day and was never seen again. It is believed by residents of the district that the young man lost his reason and has been living, like an animal, in the woods since then.

That story was tucked discreetly on an inside page, perhaps in consideration of the fact that it was no better than third hand, but on June 21 the wild man was back on the front page again:

Nanaimo, June 20—The wild man of Horne Lake has been seen again, this time by James Kincaid at Little Qualicum, who has written Government Agent Bray the following letter, which contains a somewhat startling request for permission to take a pot shot at the man on sight.

"Little Qualicum, June 18—I write you to let you know that I saw the wild man on the 18th of June, just by the Little Qualicum schoolhouse. I was out visiting and was going home. I thought it was a man, and was going to get off my bike and walk with him, and whistled for him to get out of the way. When I was about ten yards from him he turned his head and when he saw me coming he made a jump in the bush and ran away. I passed him. I was about three yards from him. He is a man about six feet high, and a very stoutly built fellow. It is the same looking thing I saw up at Horne Lake last fall when I was hunting up there. Would it be advisable to shoot him if seen again? Please let me know by return mail.

> *Yours truly.*
> *(sgd) James Kincaid"*

Bray has informed Kincaid that it is unlawful to shoot Mowglies within the province of British Columbia at any time.

On July 20 of the same year the *Colonist* had one last mention of the wild man, a paragraph on an inside page as follows:

He is Still Wild—The Cowichan Leader says: "The wild man of Vancouver Island has again been seen by a prospector while out in the mountains last week, near Cowichan lake. He reports seeing what he believes was the much-talked-of wild man. He saw something through the bush, and at first sight thought it was a bear, and raised his rifle, moving a little closer, when to his surprise a man straightened up before him. He immediately lowered his gun and shouted to him, but the wild man

at once sprang into the thicket and was soon lost to view. The prospector tried to follow his track, but on account of the dense undergrowth was forced to give up the chase."

There is an assumption in these stories that the same creature is involved in all the sightings, and an impression is created that the reports all come from the same rather limited area, but that is not actually the case. Horne Lake and Little Qualicum are close together, but Union Bay is 20 miles away, and from the head of Campbell River to Cowichan Lake would be about 100 miles as the crow flies and is very rugged country. There were no further reports from Vancouver Island in the early years of the century, but there have been several recent ones.

Idaho and Montana also have their old newspaper stories, one each. Both are known to me only from reprints in Eastern newspapers which were sent to me by Tim Church, who seems to have a constant supply of such stories but will never say how he comes by them. The Montana story is the earlier of the two, printed in the New Haven, Connecticut, *Evening Register* November 11, 1892, and quoting the Anaconda, Montana, *Standard*:

A MONTANA MONSTER
It Looks Like A Man
And Eats Bears and Sheep

Some of the old hunters and Indian fighters who are still holding out in the city should endeavor to find a wild eyed individual who came in from the mountains this morning. Whether he discovered a new brand of whiskey or whether it was the loneliness of his life in the mountains that caused him to see visions and hear sounds is not known, but, whatever the cause, he told a story that knocks Joe Klaffki's ghost story, attested to by Jack Brennan, completely in the shade.

He said that over in the range of mountains which forms a part of the Wyoming line he had seen evidence of the existence of a creature whose genus was unknown to him. He also claimed to have obtained a glimpse of the "variant," but always when he was unarmed, and as its appearance was such as not to invite close inspection he had never sought to get near enough to it to see just what it was.

He says the animal is covered with hair, but in form it is not unlike a man, a resemblance that is increased by the creature's habit of rising on its haunches and walking on its hind legs after the manner of a gorilla.

It may seem fantastic to suggest that there could be apelike creatures eight feet tall and weighing more than a thousand pounds, but actually a big male gorilla like this one would fill the bill if he had legs as long in proportion to the rest of his body as a man's are.

After having seen the animal the man said he could account for the existence of the torn and partly eaten carcasses of several large bears and also of one mountain sheep that he claimed to have found in the vicinity of where the unknown animal apparently makes its headquarters. The stranger says he will return to the mountains shortly and will pilot anybody who may desire to visit the locality to the exact spot where he last saw the monster.

The Idaho report appeared in the February 5, 1902, edition of the Wildesboro, North Carolina, *Chronicle·*

HAIRY MONSTER

According to the Pocatello, Idaho, correspondent of the *Desert News*, the residents of the little town of Chesterfield, located in an isolated portion of Bannick County, Idaho, are greatly excited over the appearance in that vicinity of an eight-foot, hair-covered human monster. He was first seen on January 14, when he appeared among a party of young people, showed fight, and, flourishing a large club and uttering a series of yells, started to attack the skaters, who managed to reach their wagons and get away in safety.

Measurements of the tracks showed the creature's feet to be 22 inches long and 7 inches broad, with the imprint of only four toes. Stockmen report having seen his tracks along the range west of the river. The people of the neighborhood, feeling unsafe while the creature is at large, have sent 20 men on its track to effect its capture.

91

Theodore Roosevelt's Story

In 1961 Ivan Sanderson published a book, *Abominable Snowmen, Legend Come to Life*, in which he told the story of reports of man-like beings all around the world. One story from which he quoted a length had been written in 1892 by Theodore Roosevelt, in a book titled *Wilderness Hunter*. There is no definite date for the events described, but it must be one of the very earliest stories from the Northwest, as well as the goriest. Here is the full text:

Frontiersmen are not, as a rule, apt to be very superstitious. They lead lives too hard and practical, and have too little imagination in things spiritual and supernatural. I have heard but few ghost-stories while living on the frontier, and those few were of a perfectly commonplace and conventional type.

But I once listened to a goblin-story which rather impressed me. It was told by a grizzled, weather beaten old mountain hunter, named Bauman, who was born and had passed all his life on the frontier. He must have believed what he said, for he could hardly repress a shudder at certain points of the tale; but he was of German ancestry, and in childhood had doubtless been saturated with all kinds of ghost and goblin lore, so that many fearsome superstitions were latent in his mind; besides, he knew well the stories told by the Indian medicine-men in their winter camps, of the snow-walkers, and the spectres, and the formless evil beings that haunt the forest depths, and dog and waylay the lonely wanderer who after nightfall passes through the regions where they lurk; and it may be that when overcome by the horror of the fate that befell his friend, and when oppressed by the awful dread of the unknown, he grew to attribute, both at the time and still more in remembrance, weird and elfin traits to what was merely some abnormally wicked and cunning wild beast; but whether this was so or not, no man can say.

When the event occurred Bauman was still a young man, and was trapping with a partner among the mountains dividing the forks of the Salmon from the head of Wisdom River. Not having had much luck, he and his partner determined to go up into a particularly wild and lonely pass through which ran a small stream said to contain many beaver. The pass had an evil reputation because the year before a solitary hunter who had wandered into it was there slain, seemingly by a wild beast, the halfeaten remains being afterwards found by some mining prospectors who had passed his camp only the night before.

The memory of this event, however, weighed very lightly with the two trappers, who were as adventurous and hardy as others of their kind. They took their two lean mountain ponies to the foot of the pass where they left them in an open beaver meadow, the rocky timber-clad ground being from there onward impracticable for horses. They then struck out on foot through the vast, gloomy forest, and in about four hours reached a little open glade where they concluded to camp, as signs of game were plenty.

There was still an hour or two of daylight left, and after building a brush lean-to and throwing down and opening their packs, they started upstream. The country was very dense and hard to travel through, as there was much down timber, although here and there the sombre woodland was broken by small glades of mountain grass. At dusk they again reached camp. The glade in which it was pitched was not many yards wide, the tall, close-set pines and firs rising round it like a wall. On one side was a little stream, beyond which rose the steep mountain slope, covered with the unbroken growth of evergreen forest.

They were surprised to find that during their absence something, apparently a bear, had visited camp, and had rummaged about among their things, scattering the contents of their packs, and in sheer wantonness destroying their lean-to. The footprints of the beast were quite plain, but at first they paid no particular heed to them, busying themselves with rebuilding the lean-to, laying out their beds and stores and lighting the fire.

While Bauman was making ready supper, it being already dark, his companion began to examine the tracks more closely, and soon took a brand from the fire to follow them up, where the intruder had walked along a game trail after leaving the camp. When the brand flickered out, he returned and took another, repeating his inspection of the footprints closely. Coming back to the fire, he stood by it a minute or two, peering out into the darkness, and suddenly remarked, "Bauman, that bear has been walking on two legs." Bauman laughed at this, but his partner insisted that he was right, and upon examining the tracks with a torch, they certainly did seem to be made by but two paws or feet. However, it was too dark to make sure. After discussing whether the footprints could possibly be those of a human being, and coming to the conclusion that they could not be, the two men rolled up in their blankets, and went to sleep under the lean-to.

At midnight Bauman was awakened by some noise, and sat up in his blankets. As he did

THEODORE ROOSEVELT'S STORY

so his nostrils were struck by a strong, wild-beast odor, and he caught the loom of a great body in the darkness at the mouth of the lean-to. Grasping his rifle, he fired at the vague, threatening shadow, but must have missed, for immediately afterwards he heard the smashing of the underwood as the thing, whatever it was, rushed off into the impenetrable blackness of the forest and the night.

After this the two men slept but little, sitting up by the rekindled fire, but they heard nothing more. In the morning they started out to look at the few traps they had set the previous evening and put out new ones. By an unspoken agreement they kept together all day, and returned to camp towards evening.

On nearing it they saw, hardly to their astonishment, that the lean-to had again been torn down. The visitor of the preceding day had returned, and in wanton malice had tossed about their camp kit and bedding, and destroyed the shanty. The ground was marked up by its tracks, and on leaving the camp it had gone along the soft earth by the brook, where the footprints were as plain as if on snow, and after a careful scrutiny of the trail, it certainly did seem as if, whatever the thing was, it had walked off on but two legs.

The men, thoroughly uneasy, gathered a great heap of dead logs and kept up a roaring fire throughout the night, one or other sitting on guard most of the time. About midnight the thing came down through the forest opposite, across the brook, and stayed there on the hillside for nearly an hour. They could hear the branches crackle as it moved about, and several times it uttered a harsh, grating, long-drawn moan, a peculiarly sinsiter sound. Yet it did not venture near the fire.

In the morning the two trappers, after discussing the strange events of the last 36 hours, decided that they would shoulder their packs and leave the valley that afternoon. They were the more ready to do this because in spite of seeing a good deal of game sign they had caught very little fur. However it was necessary first to go along the line of their traps and gather them, and this they started out to do. All the morning they kept together, picking up trap after trap, each one empty. On first leaving camp they had the disagreeable sensation of being followed. In the dense spruce thickets they occasionally heard a branch snap after they had passed; and now and then there were slight rustling noises among the small pines to one side of them

At noon they were back within a couple of miles of camp. In the high, bright sunlight their fears seemed absurd to the two armed men, accustomed as they were, through long years of lonely wandering in the wilderness, to face every kind of danger from man, brute or element. There were still three beaver traps to collect from a little pond in a wide ravine near by. Bauman volunteered to gather these and bring them in, while his companion went ahead to camp and made ready the packs.

On reaching the pond Bauman found three beavers in the traps, one of which had been pulled loose and carried into a beaver house. He took several hours in securing and preparing the beaver, and when he started homewards he marked, with some uneasiness, how low the sun was getting. As he hurried toward camp, under the tall trees, the silence and desolation of the forest weighed on him. His feet made no sound on the pine needles and the slanting sun-rays, striking through among the straight trunks, made a gray twilight in which objects at a distance glimmered indistinctly. There was nothing to break the gloomy stillness which, when there is no breeze, always broods over these sombre primeval forests.

At last he came to the edge of the little glade where the camp lay, and shouted as he approached it, but got no answer. The camp fire had gone out, though the thin blue smoke was still curling upwards.

Near it lay the packs wrapped and arranged. At first Bauman could see nobody; nor did he receive an answer to his call. Stepping forward he again shouted, and as he did so his eye fell on the body of his friend, stretched beside the trunk of a great fallen spruce. Rushing towards it the horrified trapper found that the body was still warm, but that the neck was broken, while there were four great fang marks in the throat.

The footprints of the unknown beast-creature, printed deep in the soft soil, told the whole story.

The unfortunate man, having finished his packing, had sat down on the spruce log with his face to the fire, and his back to the dense woods, to wait for his companion. While thus waiting, his monstrous assailant, which must have been lurking in the woods, waiting for a chance to catch one of the adventurers unprepared, came silently up from behind, walking with noiseless steps and seemingly still on two legs. Evidently unheard, it reached the man, and broke his neck by wrenching his head back with its fore paws, while it buried its teeth in his throat. It had not eaten the body, but apparently had romped and gambolled around

THEODORE ROOSEVELT'S STORY

it in uncouth, ferocious glee, occasionally rolling over and over it; and had then fled back into the soundless depths of the woods.

Bauman, utterly unnerved, and believing the creature with which he had to deal was something either half human or half devil, some great goblin-beast, abandoned everything but his rifle and struck off at speed down the pass, not halting until he reached the beaver meadows where the hobbled ponies were grazing. Mounting, he rode onwards through the night, until beyond the reach of pursuit.

This Kwakiutl pole shows the Dsonoqua (cannibal woman) holding her son. The equivilent of the Salish "Sasquatch", the Dsonoqua is the main crest of the Nimpkish tribe.

David Thompson's Story

Second-hand stories like that written by Theodore Roosevelt depend on the reliability of the original informant, not the author, so the Roosevelt reputation doesn't really lend any weight to that account. As to old newspaper stories, anyone familiar with Mark Twain's career knows that the papers occasionally took it upon themselves to spoof their readers, and it would not be unreasonable to assume that the "Wild Man" could have been a standard form of that exercise. Some of the old stories read as if that might be the case, while others are very matter-of-fact.

There is one very old report that is completely exempt from either of those lines of criticism. The problem with it is that it may not involve a sasquatch, but it does prove beyond question that a serious report of something very mysterious has been in circulation for more than 100 years without ever being explained and without ever becoming general knowledge.

The example in question is to be found in the writings of David Thompson, surveyor and trader for the Northwest Company, who has been called the greatest land geographer of all time. In the early years of the Nineteenth Century he explored much of what is now known as the Pacific Northwest, reaching the mouth of the Columbia only five years after Lewis and Clark. He made his strange discovery at the beginning of that particular trip, as an indirect result of some trouble with Indians.

Thompson had antagonized the powerful Blackfoot confederacy on the western plains by trading guns to their enemies the Kootenais west of the Rockies, and when he tried to set out on his usual route into the mountains via the North Saskatchewan River the Blackfeet turned him back. He then went north to the valley of the Athabaska, and January of 1811 found him struggling in deep snow and bitter cold to cross the Rockies near the present site of Jasper, Alberta, the first white man to come that way.

Thompson maintained a daily journal recounting the main features of every day of his travels. Years later he used it in the preparation for publication of his *Narrative*, a book about his travels which is also in diary form. The journal has been preserved and is now in the archives of the province of Ontario. The *Narrative* has been published twice and is a standard historical reference work available in every major library. Surely it should be safe to assume that there cannot be any reports of un-

DAVID THOMPSON'S STORY

known animals in such a volume, or it would be general knowledge.

Well, for January 7, 1811, the journal entry reads:

I saw the Track of a large Animal - has 4 large Toes abt 3 or 4 In long & a small Nail at the end of each. The Bal of his Foot sank abt 3 In deeper than his Toes - the hinder part of his Foot did not mark well. The whole is abt 14 In long by 8 In wide & very much resembles a large Bear's Track. It was in the Rivulet in abt 6 In snow.

Generally speaking, the *Narrative* is a condensation of the journal, but on this matter Thompson elaborated considerably in the later work:

January 7th. Continuing our journey in the afternoon we came on the track of a large animal, the snow about six inches deep on the ice; I measured it; four large toes each of four inches in length, to each a short claw; the ball of the foot sunk three inches lower than the toes, the hinder part of the foot did not mark well, the length fourteen inches, by eight inches in breadth, walking from north to south, and having passed about six hours. We were in no humour to follow him; the Men and Indians would have it to be a young mammoth and I held it to be the track of a large old grizzled bear; yet the shortness of the nails, the ball of the foot, and its great size was not that of a Bear, otherwise that of a very large old Bear, his claws worn away; this the Indians would not allow.

Later on in the *Narrative* he refers to the incident again:

I now recur to what I have already noticed in the early part of last winter, when proceeding up the Athabaska River to cross the Mountains, in company with....Men and four hunters, on one of the channels of the River we came to the track of a large animal, which measured fourteen inches in length by eight inches in breadth by a tape line. As the snow was about six inches in depth the track was well defined, and we could see it for a full one hundred yards from us, this animal was proceeding from north to south. We did not attempt to follow it, we had not time for it, and the Hunters, eager as they are to follow and shoot every animal made no attempt to follow this beast, for what could the balls of our fowling guns do against such an animal. Report from old times had made the head branches of this River, and the Mountains in the vicinity the abode of one, or more, very large animals, to which I never appeared to give credence; for these reports appeared to arise from that fondness for the marvellous so common to mankind: but the sight of the track of that large beast staggered me, and I often thought of it, yet never could bring myself to believe such an animal existed, but thought it might be the track of some monster Bear.

Obviously Thompson was not satisfied that what he had seen could be a bear track. He did not say that the animal was two-legged, yet he describes only one type of track. A bear would have to leave the tracks of his front paws as well, and they would be much shorter. Moreover the Indians and French voyageurs with Thompson would not accept it for a bear track.

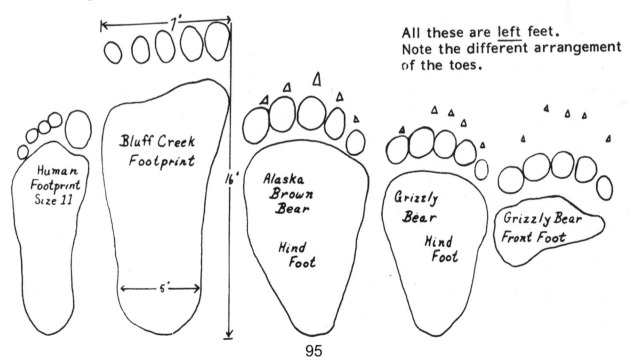

All these are <u>left</u> feet.
Note the different arrangement of the toes.

Human Footprint Size 11

Bluff Creek Footprint

Alaska Brown Bear Hind Foot

Grizzly Bear Hind Foot

Grizzly Bear Front Foot

What the Prospectors Saw

Prospectors have been responsible for a high proportion of the early sasquatch reports, and for some of the most recent ones as well. The stories from the Sixes River in Oregon; Ape Canyon in Washington, and the Albert Ostman story in British Columbia all involved prospectors. There are several more good stories from British Columbia, two of them involving unique observations, and there is also a brief but interesting story that came to light very early in the investigation and was one of those sworn to by the informant.

Following is the text of a statutory declaration sworn by Charles Flood, of New Westminster, B.C., in 1957:

I am 75 years of age and spent most of my life prospecting in the local mountains to the south of Hope, towards the American Boundary and in the Chilliwack Lake area.

In 1915, Donald McRae and Green Hicks of Agassiz B.C. and myself, from Hope, were prospecting at Green Drop Lake, 25 miles south of Hope and explored an area over an unknown divide, on the way back to Hope, near the Holy Cross Mountains.

Green Hicks, a half-breed Indian, told McRae and me a story, he claimed he had seen alligators at what he called Alligator Lake, and wild humans at what he called Cougar Lake. Out of curiosity we went with him; he had been there a week previously looking for a fur trap line. Sure enough, we saw his alligators, but they were black, twice the size of lizards in a small mud lake. (Presumably salamanders. The native lizards are very small.)

A mile further up was Cougar Lake. Several years before a fire swept over many square miles of mountains which resulted in large areas of mountain huckle-berry growth Green Hicks suddenly stopped us and drew our attention to a large, light-brown creature about eight feet high, standing on its hind legs pulling berry bushes with one hand or paw towards him and putting berries in his mouth with the other hand, or paw.

I was still wondering and McRae and Green Hicks were arguing. Hicks said, "It is a wild man," and McRae said, "It is a bear." The creature heard us and suddenly disappeared in the brush about 200 yards away.

As far as I am concerned the strange creature looked more like a human being. We seen several black and brown bear on the trip, but that thing looked altogether different. Huge brown bear are known to be in Alaska, but have never been seen in southern British Columbia.

I never have seen anything like this creature before or after this incident in 1915, in all my days of hunting and prospecting in British Columbia.

Another prospector of long experience, whom I interviewed after his years in the bush were over was Burns Yeomans, of Deroche, B.C., who told one of the most interesting stories I have ever heard. Here is part of the transcript of a tape-recorded interview:

Q–How long have you lived in this part of the country?
A–I was born in this part of the country.
Q–How old are you now?
A–Sixty-four.

Q–Have you spent much time in the bush?
A–Quite a lot of it.
Q–What sort of work would you be doing?
A–Prospecting a lot.
Q–The occasion we're interested in, what year would that be?
A–I'm not sure, 1939 or 1940, one or the other.
Q–What time of year?
A–August.
Q–Were you alone.
A–No. Actually there were five of us up there, but there were only two of us at the top of the hill where we could see the other side.
Q–What were you doing up there then?
A–Looking for the molybdenite that's supposed to be in that country.
Q–Where did you start from? You were up Harrison Lake?
A–Yes. We were up on Silver Creek, the headwaters of Silver Creek.
Q–About how far.
A–Well, I imagine about 14 miles.
Q–And then where did you go?
A–We went east up that mountain...it's straight east from that mine.
(There was a cabin that had been built the year before near a molybdinum prospect.)
Q–Do you know how high it would be?
A–I don't...but it was in August and there was still snow on the top of that ridge.
Q–When you got to the peak you could see down on the other side?
A–On the other side down in a big valley there.
Q–What was the valley like?
A–Well, it's fairly open.
Q–Any idea how high it would be?
A–No. I imagine it would be 5,000 feet anyway.
Q–The mountain would be a good deal higher than that?
A–Yes.
Q–What did you see in the valley?

WHAT THE PROSPECTORS SAW

A-We saw these animals. I don't know what they were but there were four or five of them, wrestling just like men, down in this valley.

Q-What color were they?

A-They were about three quarters of a mile away and we didn't have any glasses with us, but they looked to me to be black. They were a dark color anyway.

Q-Could you tell whether they had hair.

A-Yeah, well I don't know. The color of them, it looked like hair.

Q-It didn't look like skin?

A-No.

Q-It didn't look like clothes?

A-No.

Q-They were wrestling like people?

A-Just like men would do.

Q-Having fun, you'd say?

A-Yes.

Q-Did you see them stand up on two legs?

A-Yes they were. I can't recall them getting around any other way but on two legs. As far as I remember of it they were on two legs all the time.

Q-Is there any possibility that they could have been bears?

A-Well, I don't know, but if there's any sasquatches in this country, I seen them. I didn't think they were bears.

Q-Could you give us an idea how long you were there?

A-About half an hour.

Q-You were sitting up there for about half an hour, watching four or five black creatures that looked like men wrestling? They kept wrestling all that time?

A-Seemed to, yes.

Q-Any sign of them getting hurt at all?

A-No, it didn't seem to. One would throw the other down, he'd jump right up on his hind legs again.

Q-Did they grab each other round the neck?

A-Well I don't remember that. I don't remember just how they done it, but it seemed like a wrestling match.

Q-Did you have an impression when you were watching them of what size they were?

A-I'd say they were seven feet high anyway.

Q-Bigger than a man?

A-Bigger than a man, yeah, I'd say they'd weigh 400 pounds or more.

Q-You think they'd be built heavier than a man?

a-Yeah....mind you with hair on....but they looked to me they'd weigh that much anyway.

Q-Had you ever been there before.

A-No.

Q-Did you ever go again?

A-No, I've never been back there.

Q-Ever seen anything else like this anywhere else?

A-Never anywhere else in this country. I've been all over these mountains.

The other story was first reported to the British Columbia Museum after an appeal was made for information following the Blue Creek Mountain tracks and the Patterson movie. A creature, and more than one set of tracks were seen by two prospectors, and both a report of the sighting and a sketch of the creature and the tracks were entered in their field notes. The drawings are of particular value, since all of them, including that of the creature itself, were made while the object being sketched was in view and under study.

The work of a prospector involves considerable secrecy, and reliability of his reports is an essential element in his continuing in the occupation. Thus a prospector who works for wages often cannot say where he has been, and he certainly is not going to say he saw a sasquatch while he was there. For those reasons I am not able to identify either the people or the location—in fact I have been given the location only in general terms.

It was during the last week of June, 1965, at about the 4,000 foot level in a valley somewhere northwest of Pitt Lake in British Columbia, that the two men came upon a set of tracks in the snow. The tracks were recent. The toes showed clearly, and there were only four of them. Each print measured two boot lengths from toe to heel and one boot length across—about 24 inches long and 12 inches wide at the base of the toes. The impressions were "as flat as plywood," with no slide in or kick out. The snow in the bottom of the prints had a pink tint.

The two men followed the tracks up a valley until they led to a small lake which was still covered with ice. In the ice a large hole had been broken, with pieces of the broken ice lifted out beside the hole. All along where the tracks went there were marks of something being dragged. Left and right footprints were well separated from side to side and between them the snow had been scraped by something wide but not very heavy. Outside the line of the prints there were deep grooves, as deep as the footprints.

Unable to fathom the meaning of these marks in the snow, the men started on around the little lake. Then they saw across the lake a creature watching them. It was about as far uphill as they were and only "a

97

WHAT THE PROSPECTORS SAW

stone's throw away." It stood upright, like an enormous man, and was auburn in color except for the hands, which looked yellow. The two men stopped and stared at the figure, which stayed immobile. They sat down and had a cigarette and a chocolate bar, and one of the men drew a sketch. The creature had a head resting very close to square-set shoulders and appeared to be very wide. Its legs were together. The features seemed flat, but were not distinct. Arms hung down in front of it, reaching below the knees, and they swayed slowly from side to side, just a little, as if the creature might be shifting its weight from one foot to the other. Fingers were held together and the hands looked the shape and size of canoe paddles. The hair that covered it everywhere seemed longer on the head and perhaps thinner on the arms. It was noon and the sun was shining.

The men discussed estimates of the creature's height, using trees near to it as a means of comparison; one guessing it at 10 to 12 feet, the other 12 to 14. It could even be as much as 15 feet, they decided later. They had a fairly accurate measuring stick in the trees. Conifers put out one set of branches each year, marking a year's growth, and they could compare the trees beside the creature with those on their own side of the lake.

The creature did not move, so they went on their way. Later in the day they returned by the same route. There were more tracks around but the creature was gone.

The following day they left the valley, climbing over a ridge, and on a plateau among some small pothole lakes they saw more tracks, but only about three quarters the size of the others. These were old impressions, the compacted snow in the bottom of the tracks had resisted melting and was

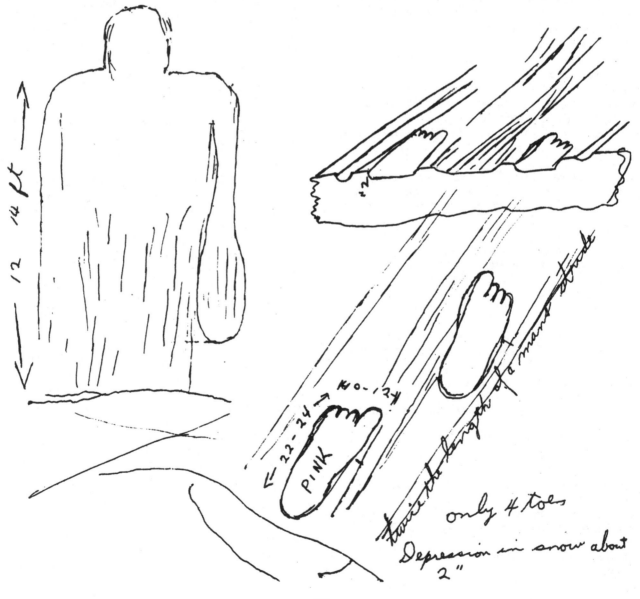

12 14 ft

← 22-24 →

140-124

PINK

twice the length of a mans stride

only 4 toes

Depression in snow about 2"

98

WHAT THE PROSPECTORS SAW

standing above the level of the snow around them, but they could still tell the front and back of the footprints. The trail led to and from one of the potholes on which it appeared that the snow had been pushed back and a hole more than five feet across made in the ice. Beside the pothole but not out on the ice were much smaller tracks, only about 10 inches long.

A few days later one of the men went back to the scene in a helicopter with a newspaper reporter. They photographed the big tracks in the valley, but by this time they had lost much of their shape. Then on the way out they saw a fresh set of big tracks on a ridge but could not land there to inspect them. Following them they came to a cliff, with the tracks leading right to the edge. There was no snow at the bottom.

The men intended to go back again when conditions at that altitude were the same, but have never been working in that area at that time of year. They have no reason to be puzzled about what makes the tracks, since they saw one of the creatures, but they are at a loss to explain the drag marks or the holes in the ice.

At the time we first heard it that story seemed to promise an easy way to advance the sasquatch investigation, since holes in the ice on mountain lakes would be easy to locate from the air, but the promise has not been realized. Such holes would be easy to see alright, as I have verified for myself, except that we could never find any. The amount of flying done has been limited, as it is very expensive, but it was certainly enough to establish that very few tracks of any sort are made in the high snowfields and there is nothing inhabiting the area that makes a regular practice of opening holes in the ice.

The mountain country around the head of Pitt Lake is extremely rugged and quite a few people have gone in there and never come out. It is supposed to hold a lost gold lode of fabulous wealth, which is why some of the people have gone there, but whether or not the story of the gold is true the story of the missing people certainly is. I have noted the gradual increase in the total during my years in the newspaper business. The terrain itself provides plenty of reasons why lone venturers might never be seen again, but there are persistent traditions that the sasquatch have something to do with it. I do not know of anything specific to support that tradition, however, and besides the prospectors' story I only have one sighting report from the Pitt Lake area on record. It reached me as a result of publication of the first of my books in 1968.

John Rodgers, a columnist for the Vancouver *Sun,* reviewed the book and received the following letter, dated December 15, 1968:

Dear John;

Very interested in your article this date on the Susquatch and I will certainly get a copy of John Green's booklet now published.

Here is the story. Thirty-five years ago there was a stock broker office on Dunsmuir operated by Cartwright and Crickmere. Cartie had a cabin cruiser, was a bachelor, hard as nails in business and with a heart of gold for those he liked, and an experienced outdoorsman. One week-end he asked my wife and I along on a party of eight to go to the head of Pitt Lake. I was an ardent rockhound, and this was virgin territory for me to possibly add to my collection, and Cartie wanted to run down a clue he had as to some lost mining prospect.

In the morning we left the rest of the party to amuse themselves for the day and Cartie and I climbed some fifteen hundred feet and rested at the edge of a small plateau to eat our lunch. We had our haversacks and small hammers but were otherwise unarmed.

A movement behind a thicket some quarter mile away caught my eye and I said, "Cartie, there is something down there." He looked and then asked for the field glasses. We both thought it was a black bear feeding on berries, then he exclaimed, "Here, look at its face!" Through the glasses it was quite plain—a human face on a fur-clad body. "What the hell," I said," he must be a hermit of some kind, but look at the size of him." Cartie replied, "wait until he leaves and let us go down and look at the tracks."

So we waited. I don't think the creature saw us, though he or she may have sensed us, as presently it went away across the plateau and vanished among the rocks. We went down after a suitable interval and examined the tracks, which were quite distinct. Cartie looked pretty grim and said, "Let's go back. What you have just seen is a Susquatch; don't mention this to anyone, not even to your wife. No one will believe you, you will just be laughed at and you will have a miserable time of it! Just forget the whole thing and keep quiet."

So I did, and I have, until now.

The Logger's Story

By no means all of the people interested in the sasquatch spend their time interviewing witnesses or digging up old stories. Some go hunting for the animal itself. Of course the odds against them are long, but if they succeed they have the satisfaction of experiencing something for themselves. They may be no more successful than any other witness in convincing other people, but they know what they have seen.

One man who spent many solitary hours in searching had phenomenal success and has contributed a great deal of information, yet is unknown to most of the public. He has passed on what he has seen, so that it is on the record, but he has asked to remain unidentified and those he has dealt with have generally respected his wishes.

His motivation to spend as much time as he did was strong and understandable, since it began with an accidental sighting that is one of the most unusual on record—the only close observation of a "family" group involving an infant. After that he spent a great deal of time in trying to see more, and at first was remarkably lucky, but later spent years having no better luck than the rest of us.

All of his experiences took place in the mountainous area drained by the Molalla and Clackamus Rivers south of Estacada, Oregon, where he worked as a logger. His first, accidental, sighting was in October, 1967, the same month that Roger Patterson and Bob Gimlin got the movie. After that dedicated searching resulted in further sightings in the spring of 1968 and in November and December of that year. From then until a heart attack restricted his activities several years later he continued to search without another sighting, although he did find some very interesting tracks in February of 1969.

As far as I know the report of his first encounter is the only one ever put on tape. This is the story he told:

I was supposed to be watching a cat skinner as he was fire trailing, but it was awful cold, and I walked a mile or so down the trail, because he had no need of anyone at that time, and I thought I'd warm up and see the country. Up where he was it was a cold east wind blowing; a little farther down it was a west wind coming in. It was late fall, the last weekend of the deer season, I think, in 1967.

It was just a mountain trail—they have several of them up there—footpaths, and for horses. The elevation was about between four and five thousand feet. I came out

lower down, into the fog, before I saw anything, and the fog was freezing on the trees because it was so cold, but if the wind would blow, the fog would break, and fall off. That made it kind of noisy, it sounded like walking.

I came around a bend—well, first I noticed some rocks that were turned over. All the other rocks were wet, because of the fog, but these rocks were dry. Then I looked up, about forty or fifty feet, up on a ridge of rock, and I saw these animals there—looked like human or just about. Large male; the female wasn't so large; and a small baby— well not really small; it was moving with them. It was standing up, mostly. The two older ones were squatting down and sort of bending, as they picked up rocks and smelled them. They were kind of careful. They moved on for a few minutes, and then finally the male found possibly what he was looking for and dug real fast down into the rocks, which were large boulders—not the round type of rock, but the flat, sharp kind.

I could not explain why those rocks were there; there hadn't been a slide or anything. They were on top of the ridge, so they couldn't have come down from anywhere. They are loose, quite a few holes underneath them, and they are as if they had been broken up—definitely not the round river-type rock. But they (the animals) would pick them up, and, after they smelled them, they would lay them down, on top of each other. They didn't just lay them back where they had picked them up, they stacked them up, in piles. And when the male found what he was looking for he really made the rocks fly. The big rocks weighed fifty, sixty or even possibly a hundred pounds; he just jerked them out with his hand. He didn't seem to take any precautions for his safety. Later on I looked, and there was some rock there that could have fallen on him, but he wasn't concerned.

He brought out what appeared to be a grass nest, possibly some stored hay that small rodents had stored there. He dug through that, and brought out the rodents. It seems they ate them. The rodents appeared to be in hibernation, or asleep, or something. There were about six or eight rodents. The small animal, I noticed, only got one, but the others got two or three apiece.

But about that time they became aware of my presence and well, just became alert. I was alongside of this trail that follows the ridge. I didn't remember getting there, but I was squatting down behind a small tree when I became aware of where I was. As soon as they realized I was there they sud-

100

Jim Green (5' 10" tall) standing in the hole the sasquatch dug. Note the piles of rocks on the skyline. There are other piles in the picture, but they are hard to see.

denly began to move, real quiet, behind some low-hanging limbs on a tree there. I didn't see them again after that.

I tried to follow their tracks in the direction I thought they would have to go, but I couldn't find any, although there was heavy frost there. But the next day I found two tracks, a heel print and the front part of the foot, the toes, but they were in a different direction—the direction from which I had come—and I never did connect them up with exactly which direction they had gone, or know anything about them

The footprints, I would say, were about twelve to fifteen inches, but there wasn't enough of the track to tell exactly. They were possibly five inches wide, I don't know, at the widest part. I don't think they could have been six. I didn't know if it was one of the animals I had seen that made the footprints.

I saw the toe print as I came up out of the old landing. I saw the heel print as I went in. The heel print gave me the impression that the heel protruded. The tracks were in dirt. It was just as if you had a level piece and scooped it out for about two feet deep, and it would cave in some. It (the animal) had stepped down into that, and left a heel print, and as it stepped out on the other side you could see the toe print.

When I left the catskinner, he was on Low Creek, but I had walked to Jim's Meadows, possibly a mile or more. I saw the footprints between where the catskinner was and where I had seen the animals.

After the animals disappeared I watched and looked a few minutes and then decided I didn't want to go in that direction. So I just headed back. I didn't tell the catskinner about seeing them. I didn't tell anyone about it until......asked me to ask among my crews, maybe some of them had seen them. That was the only time I ever mentioned it to any of the fellows out there, because I didn't want everybody to think I was a nut or something or other.

The only time I saw their faces was when they became alert. They gave me the impression of having a face a little like a cat, without the ears. I couldn't remember seeing any ears. It seemed like the nose was much flatter—it didn't stick out like a human's. The upper lip was very short, and seemed to be thin. I couldn't remember that it had a chin like a human has. So somehow or other I felt that it was a face more like a cat than a human.

The male was darker than the female, a dirty brown, where the female was a buckskin or fawn-colored animal. The male had much longer hair on his shoulder, head and neck, and hung in strings, like you see on an Angora goat. He was much heavier in the

THE LOGGER'S STORY

shoulders than the female. From just above the hips the male got larger; he had a very wide "small" of the back. From there on up he just got bigger and bigger. They had very rounded or stooped shoulders. The head was set lower on the shoulders than a human's. They don't seem to have a neck "stand up" as we do.

Most of the time they were not standing, but were squatting down and leaning forward to pick up the rocks. I didn't see them actually erect until they became alert that I was there. I didn't see them walk, as such. The only movement I saw was when they made a quick, short dash to get behind the limbs of the trees. I saw them move all right, but in a humped-up, stooped-over position, just moving across the rocks. But they were upright when they made that quick dash at the end. It seemed to me that the mother picked the baby up on her lap, and ran holding the baby in front of her, possibly right below the breast, and her breasts hung real low, much lower than on a human.

I couldn't say how thick through the body these animals were, but they were very heavy-set—particularly thick and heavy at the small of the back, and on up through the ribs. I think the male was over six feet tall, but I'm an awful judge of height or weight or anything. I didn't think the female was quite as tall as the male, in fact I think she came possibly up to his shoulder, but I saw them standing up so little I don't know. But they were much larger than a human, much bulkier. The baby didn't come up to the mother's hips, actually, I don't think, but I don't remember for sure. The first time I saw them standing up was as the male stepped out of the hole he dug with the grass, but it was only a very short while until they took off. I didn't see them other than that.

Q-How did they eat?

A-They ate just taking it in their hand and eating it as one of us would if we were eating a banana. They ate it skin, feathers and all—just bit it in two, and as they would bite part of it well then just cram the other right on in. The little one, though, he had a little more difficulty, because he didn't have quite enough room for all of it, where the older ones did. It wasn't like a human would hand the food to the baby, he had to get his—he was scratching through the grass that they had and got it himself, and the female did the same thing. They gave you the impression in that way of not taking care of the baby like people would. I've been wondering now if that group lived together as a family, and I hope to go back and look

into it deeper.

Q-Did you form any impression of the proportions of, say, the legs in relation to the rest of the height? Would they be like a long-legged man, or short legged?

A-I don't know; I couldn't say for sure; but the arms were such that when they squat down they have to bend forward to pick up anything—their arms are not long enough to reach. This one that was digging just seemed to go right on down. I didn't remember seeing him get up, but as he was down here he was just digging and kept on going down, and....well at that time I couldn't see exactly where he was, because I was down, and they were a little bit on the side of this rock, which kind of levels off some, and he went down, and so I could not see exactly what he was doing in there, but I did see when he came out. At that time I was a little bit nervous....I'm not sure, now, about half of it—seemed like a bad dream for a while. I just couldn't believe it was really happening. It just couldn't be, but it is.

Q-Did you notice the hands at all?

A-I noticed that it had hands. I did not notice if it had thumbs. I couldn't tell from the way it worked—it didn't seem to use the thumb. And I didn't see any ears. I didn't see any knees projecting when it squatted. They were in an awkward position because of the rocks, and they couldn't just squat down like we would on a floor. They would be on different levels, and off too far to be comfortable. That's as close as I could explain it.

When they went from place to place they would shift in position, according to the terrain. The male, well, actually both of them, seemed to be moving in a certain direction, possibly from tracing the small rodents. I thought possibly it was the scent left by the rodents coming up through the rocks, because it was not a runway that they could have been picking, because they were just picking the rocks up anyplace, and as they picked it up, they'd turn it over and smell it, and then lay it on the stack. They left it very definitely in a pile. They would leave anywhere from three to fifteen or twenty in one pile, as they would reach back, and then, oh, six of eight feet farther, they would leave another pile—they would start laying them in another pile.

With Rene and my daughter Kathryn and son Jim I went with this man to the spot where he had seen the three creatures. We found piles of rocks just as he said, not only at the spot he showed us but on almost every area of broken rock we came upon in a couple of hours of scrambling around the

THE LOGGER'S STORY

mountain. They were obviously piles manufactured by someone or something, there was no way the rock could have come to rest in that way naturally, and there were dozens of them.

The hole he saw the male sasquatch dig was about five feet deep and almost as steep-sided as a well. No bear or anything else without hands could have lifted out the rocks. A man could undoubtedly figure out a way to do it if he had a reason to, and could also have made all the rock piles, but it would have been a great deal of work for no apparent purpose.

At that time the logger had already had his second experience with a sasquatch, and he showed us that site also. It was a good deal lower down, quite close to a logging road. He was not actually searching, but had stopped at what he thought would be a good place for some shooting to sight in a rifle, when he saw a dark brown animal five or six feet high and very heavily built standing eating leaves from a willow bush in an overgrown clearing. It was about 100 feet from him. He noted that it had breasts, again located somewhat lower than human breasts, but not hanging; and that its thumb was short and set farther back on the hand than a human thumb. It was using its hands to strip the leaves from the bush and stuff them in its mouth, but after a few seconds it noticed him and fled into cover.

He said that he had noticed that the animals on the rock pile did not use their thumbs in picking up the rocks, but gripped them between the fingers and the palm of the hand, and he wondered if the thumbs were any use at all.

In another conversation he mentioned that the small creature had seemed to be afraid of the male and had always kept on the other side of the female away from him.

For the next half year he hunted without success, but in November he again hit the jackpot. Crossing a ridge at about 3,500 feet he found two sets of tracks, both about 16 inches long, and followed them for several miles in the snow, going down a logging road. They were not always on the road but kept coming back to it. Eventually they reached a level where the snow was petering out and he lost them in the woods. Next day he went back to where he had left off, and while casting about for more tracks he saw something dark on the snow across a small open area.

Using binoculars at a distance of less than 200 yards, he found himself looking at two sasquatches sleeping in the open, with

their backs to the sky and their knees and elbows drawn in under their bodies. I have since read that convicts put in solitary confinement in Alcatraz without bedding slept in exactly that position to minimize the loss of heat on the cold floor.

He settled down for a long vigil, and they slept for about an hour, with very little movement, then one got up, and then the other. They went to a creek a few feet away and began pulling up and eating water plants. Both were obviously females, with more pendulous breasts than the ones he had seen before, and one appeared to have a swelling in the genital area and kept rubbing itself. It also gave an occasional loud call, which he described in a computer questionnaire as "like a scream in an echo chamber." He also saw one defacate in the creek. It stepped up on a wide, low stump that was in the water, "bent forward about 45 degrees with its knees slightly bent and let fly." It then wiped itself with one hand, and then licked the hand briefly.

After feeding for about half an hour, working their way up the creek away from him, they lay down again in a new location for about an hour, then both got up and crossed the road. The one with the swelling climbed a few feet up in a dead yew tree and wailed, then both moved off into the timber above the road.

He estimated them to be only about six feet high, but very heavy. They were both dark brown, covered with shaggy, dirty hair. When they lay down they did not seek shelter, although there were trees nearby and at one time snow was falling. After they left the observer went home. He said he was "too spooked" to go over where they had been. Next day there was new snow.

His final encounter took place only about a month later. He was following some elk tracks, walking on top of old, deep snow, when he happened to look back and saw, right behind him only about 10 feet away, a scruffy looking, dark brown sasquatch at least nine feet tall. Its arms were raised in what he took to be a threatening manner, and he scrambled to try to get a revolver out from under his rain gear. However the animal quickly ran and ducked behind the roots of a blown-down tree. He himself "left the area running" as the computer form notes. His impressions of that individual are pretty spotty, but he realized afterwards that there had been a mild smell "like an old outhouse," and that it had a long upper lip that fluttered as it blew through its mouth. He also noticed that its hands were very large and long, but with the thumb "not up where a human thumb is."

THE LOGGER'S STORY

The last thing he found, in February, 1969, was the tracks of two creatures, one with a 14-inch foot, the other 11-inch, that came out of the forest into a field where it appeared they had been eating the lower stems of clumps of grass. The tracks were made at night, for several nights, but then he found them taking off across country, entering an area where there was no snow. He then back-tracked into the woods and found that the animals had been sleeping just a short distance in out of sight. What particularly interested him was that the two tracks were always near each other but never right together.

There is a wealth of information in these observations. They depict creatures that do not have an effective opposable thumb, or if they do don't use it much. There are three accounts of sasquatches actually eating something—rock rodents, willow leaves and water weeds—and circumstantial evidence that they eat grass stems. The observation of the infant avoiding the male suggests that the trio was probably not a family group and the tracks in the field suggest the possibility of a juvenile hanging around its mother but not welcome. The other pair was made up of two adult females, again suggesting that family groups may not be the norm. Female chimpanzees have to put up with large sexual swellings when in heat, but other apes do not, and there is no other report of a sasquatch in that condition. That, plus the calling, indicates reproductive arrangements unlike those of any other ape but perhaps suitable for a species very thinly distributed. On that point, however, note that he describes a total of seven individuals all different, and tracks that appear to belong to two more, all in an area of about 200 square miles.

It is always doubtful whether any story is true, and when one person has several stories extra caution is indicated, but this man gained nothing, not even notoriety, and there were other people who confirmed that he had taken to spending all his free time in the woods. Also there were other reports from that area around that time, and some tracks photographed in snow on a road.

There was also a very good report from the same area three years after he made his last discovery. It was contained in a letter I received in July, 1973, from a man living in Oregon City. Here is what he had to say:

The encounter took place in the following locality—along the Collowash River, a small tributary of the Clackamas River, about 40 miles above the town of Estacada. This region is extremely rugged, heavily timbered and very seldom visited by man.

The sighting took place about six weeks ago. I and my nephew decided to go on a weekend fishing trip to this locale. I have fished the Collowash for 12 years, and it is one of the most beautiful, totally unspoiled places I have ever been.

We had fished all day Saturday, and were very happy to return to our camp early that evening. We proceeded to build a fire circle on a large gravel bank close to the stream. We built a fire, prepared our meal, and settled back for a much needed rest and conversation about the day's activities.

My nephew proceeded to fall asleep rather early, but I remained awake until fairly late, perhaps 11:00 p.m., feeding the fire, and feeling quite content and at peace with the world.

This peaceful attitude was rather quickly shattered. First of all, I heard a sound exactly as if someone was walking along the gravel bank close to the river. I could distinctly hear the crunch and grinding of each step.

I strained to see who or what could be making such a noise, and at that time a very large form came within view of the fire light. It was walking erect, with a very determined gait, almost fluid in motion. The creature swung its arms as it walked, almost as a man would, but slightly more pronounced. As it passed directly in front of the fire, about 20 yards away from me, it paused slightly, and gave almost an indifferent glance towards me and towards the fire.

Needless to say I was virtually thunderstruck. I rose slowly, turned my flashlight on the creature, drew my .22 pistol and fired three shots. Upon my execution of this action the creature emitted a high-pitched scream, and moved with really astounding speed over a huge log, and into the heavy timber.

I was so upset that I could not sleep the rest of the night. Early the next morning my nephew and I searched the bank for signs or tracks but found only a few depressions in the coarse gravel which could have been footprints. We also examined the point at which the creature moved into the timber. There we found no trace of its movement. The log which the creature literally stepped over was waist high and I had some difficulty climbing over it.

In a later letter he noted that the animal had a "truly nauseating acrid odor," and that it was breathing very heavily, almost wheezing, like someone with asthma or emphysema, although it was not exerting itself at all.

Gentle Giants

The gorilla enjoyed a century or so of terrorizing people all over the world by reputation and appearance. Then zoologists started to study him in his native haunts and discovered that he was a placid fellow most of the time and that when he did throw a tantrum it was mostly bluff.

It is the unknown that terrifies, and the sasquatch, never having been much bothered by prying zoologists, have remained terrifying. Not everyone considers them fearsome, but most people do, and will panic when confronted by one. Even some of the people who are looking for sasquatches but are very outspoken against shooting them go armed to the teeth, apprehensive and on the alert at all times. I would not deny that the first few times I was alone at night in the Bluff Creek area after seeing the big footprints it didn't take much to make the hair stand up on the back of my neck. Even today if I were to wake up to find a hairy giant rocking my minibus back and forth I would be unlikely to remain calm and detached. That would be an emotional reaction, however, and would not make any sense, because the sasquatch is not really unknown and should not be terrifying.

As with a lot of other matters the thing to remember here is that while we can't prove that the sasquatch exist, if they do exist we have assembled a lot of information about them. Each year there are reports of people being mauled and sometimes killed by bears. Each year there are no reports of people being mauled or killed by sasquatches. It cannot be reasonable, therefor, to consider a sasquatch to be as dangerous as a bear, yet very few people fear bears enough so they will not walk in areas where bears are known to be, and most people on seeing a bear will not feel impelled to flee.

Man is a very weak animal physically, and without a weapon couldn't win a fight against most animals half his size. The reason that not just men, but women and children, can safely roam most wild areas of North America unarmed is that the various animals that are large enough to harm them simply don't do so. Surely the sasquatch should be judged like the others, by his record rather than by his looks.

An example can be found in an experience of Clarence Fox, who was a sasquatch investigator long before even Rene Dahinden got involved, although we didn't know it until years later. In the early 1950's, when he was a schoolteacher at Wenatchee, Washington, he set out with a friend to hike down the valley of the Napeequa River, high in the Cascades. With precipitous sides and extremely dense undergrowth, the valley was considered to be impassable, and they had a hard time to get through. Right in the middle, however, they woke up one morning to find bipedal tracks, fresh, and larger than grizzly tracks, crossing the gravel bar where they had slept undisturbed. Another researcher, Keith Soesbe, in Oregon, learned of two occasions when people saw sasquatch approach sleeping men without touching them.

I woke up once on Mount St. Helens to find a herd of elk browsing a few feet away from me. It was a fascinating experience. Why should it be any different to wake up and see a sasquatch nearby? They aren't any bigger and they haven't shown any greater inclination to be dangerous.

I am not making these statements off the top of my head. It is my practice to do a thorough analysis of the information in my files every few years, and from more than 1,600 reports the last time I checked every instance of behavior even suggesting any aggressive tendency could be listed on a couple of sheets of writing paper. It is true that many Indian traditions describe the creatures as man eaters, and I don't contend that they never were, but there is certainly no indication that they are any danger to man today. The only substantial account that I have of a human being killed is in the story told by Theodore Roosevelt, and that would have been well over 100 years ago. One vicious individual in an animal population in a century is certainly nothing to make a case out of.

The one form of activity that could be considered aggression is chasing people, whether on foot, horseback, bicycle or in a vehicle. People on foot have been reported chased on 17 occasions that I have record of, mainly in British Columbia and Washington, and there have been a couple chased on bicycles and one on horseback. But in no case did the sasquatch catch up, and in five other instances where people were rushed at but did not flee the animal always backed off. There has also been some rock throwing blamed on the sasquatch but it never seems to hurt anyone.

The idea of an animal shaking a car with a person in it takes a bit of getting used to, but sasquatches have been reported to do that, and have also, apparently, tried to shake houses. Here is a fairly typical story, from the Blue Lake *Advocate*, September 24, 1964. Blue Lake is in northwestern California about 40 miles south of Bluff Creek and close to the coast:

GENTLE GIANTS

An abnormal brown bear (some call it "Bigfoot" or the Abominable Snowman), was reported seen at 1:00 a.m., Sunday, September 13th by Benjamin Wilder who told of being awakened while sleeping in his car at his camp on the water pipeline which is at four thousand foot elevation.

Wilder told how he was awakened by his car being shook up two different times. He said at first he thought it was an earthquake, but thought, with such a hard shake, there should be rocks rolling. Then, after the second time, Wilder put on his flashlight and to his surprise there was a large animal standing beside the driver's door of the car.

The animal had its two arms on top of the small car. He said it had long shaggy hair, about three inches long, on the chest. Wilder said he tried to scare it by shouting, but the bear did not move. It only made noises like a hog. Then Wilder honked the car horn. This scared the big animal and it took off, walking on its hind feet, and disappeared over the hill. Wilder said the animal never got down on its four feet and he never got to see the face of the animal.

There is also the matter of people being picked up. Albert Ostman is the only person I know of who claims to have been carried anywhere. Bill Cole may have been carried, as his hunting partner claims, but if so he was just accidentally swept up in a collision. The many stories of Indian maidens being carried off have never produced a single victim or even an eye witness. There was a young lady in Wilsonville, Oregon, who told me she had been picked up, but she wasn't carried off, she was thrown away, landing on the seat of her jeans in a patch of thistles.

It happened on August 29, 1970, at about 5 p.m., and thanks to a call from one of her neighbours I was there to talk to her the following day. She said she had heard some shooting over beyond a patch of woods on their farm and had gone down from the house across the field with the intention of ordering whoever was shooting off the property, taking her own shotgun with her. The woods were separated from the field by a barbed wire fence. She told me:

I put my gun up in the corner and started to go through the fence. I went through the fence and it just grabbed me and threw me back over the fence. I was still kind of crouched. I'm sure I didn't even have time enough to get straightened up from going through the fence.

Q–How did it pick you up?

A–I'm not sure. I just know I was flying through the air.

Q–And what did you see?

A–Well, it had hairy....I don't think it had hair on its palms....it was just hairy, and then afterwards I noticed an odor, a terrible odor.

Q–When did you see the hand?

A–When it grabbed me.

Q–Do you remember where on you the hands were?

A–I saw it on my arm.

Q–Did you see the face at all?

A–I don't think so. It was just big and hairy and threw me, and by the time I got through rolling and back up, I don't remember seeing it again. I don't know if it went right back in the bush or what happened.

Q–Did you see its eyes?

A–I thought I did. I thought they were beady, but I'm not sure if it was its eyes or what it was. They were reddish, what it was I saw.

Q–Did you see anything to indicate how big it was.

A–I thought when it threw me it stepped over the fence. It straddled the fence.

Q–At the same time it was throwing you?

A–Yes, it straddled the fence, and by the time I got through rolling and stuff it was gone.

Q–Could you see that it was something upright?

A–Yes.

Q–You never really saw the whole outline of the thing?

A–No. I know it had to be big though, to throw me.

Between the distance she was thrown and the scrambling she did before regaining her feet she was perhaps 10 or 15 yards from the fence, and she walked back to get the shotgun. When I talked to her she could not understand how she could have done that. She wasn't very keen on going back down to the end of the field even with several armed men with her. As she was starting back to the house her husband came running toward her. He said he had heard her scream. She remembered trying to scream but thought she had been unable to make any noise. Aside from being stuck by some thistles she was not hurt.

The small patch of woods certainly was not the abode of any large animal, but it was strategically placed for a daytime stop for a wild creature crossing between the Coast Range and the Cascades. It seemed possible than the animal had moved to the edge of the woods to get away from the earlier shooting and considered itself threatened when a person with a gun approached from

GENTLE GIANTS

the opposite side. What it did certainly got across the message that it did not want any company, but to have done it so swiftly yet without causing the slightest injury it must have been deliberately very gentle. The neighbour who phoned me said the bad smell of the creature was obvious on the sweater the lady was wearing. However it took me half a day to get there and by then there was no smell that I could discern.

As to what appears to be bluffing behavior with no harm done, there is an excellent example in a story told by two young couples from Yakima who went shopping in Seattle in June, 1970, and didn't want to spend the money for hotel rooms. They drove back up the road toward Yakima until they got away from civilization and then parked on a side road in the mountain forest and got out their sleeping bags. It was about 11:30 p.m. Here are excerpts from a tape recording of one of the young men talking to Roger Patterson:

My friend started talking about different calls, you know. He said he was from Missouri, said he could make a black panther sound, like a woman's scream, he said. So he did that, and I did a wolf call. About that time, after we completed our calls, we first noticed a big form on top of the hill... To me it looked a good nine feet tall, very very broad. I didn't get a good look at it...

We thought we got a glimpse of it coming down to the left of us and that was the last time we seen it for a while. About half an hour later it was coming down the road. N— shone the light on it. He started screaming "Look! look! there it is!" and it was coming down the road and I turned around and saw it.

He said he was going to stop it by blinding it in the eyes and it stopped cold and jumped behind the embankment, behind some bushes, and it looked like it was going to stalk us. We could see its hair and the outline of it and the eyes. You know, like you'll shine another animal in the eyes, in the light they'll glow.

I jumped in the car and accidentally locked the door on N— and the thing got up and rushed us and poor N— couldn't get in because the door was locked. He started screaming and everything else so I got the door unlocked and he jumped in the car and all of a sudden that thing just hit the car and really made a ruckus.

It seemed to want to get in for some reason, or maybe it was trying to be friendly or curious. Whatever it was, we didn't like it. So then it went away and we decided,

well, we're going to split, take off, but he couldn't find his keys. He figured he left them in the trunk. So we got out and sneak around there and no keys.

Then we heard....saw the thing so we think we better get back in the car, it would be safer'n out there, so we kept hearing this thing, kept seeing it and we decided well we've got to get out of here. We decided the keys have got to be around there, so we both got out and I took a pop bottle and I busted it as a weapon, but I busted it all to pieces so we stood there and here the thing came and we never got nothing....had no chance of leaving 'cause we didn't have the keys and couldn't find them.

So the thing started rushing around.... just walking around us out of curiosity.... well it didn't do any more damage after that, didn't attack the car and we just stayed there all night.

Next morning, it was about eight o'clock, we got out of the car and walked around to make sure the thing wasn't there. Actually we looked around first, we figured it wouldn't show in the daytime. We found the track, down in the cut we noticed these big tracks of this here particular thing, and they was good sized, they looked like a human's footprints.

We followed the tracks down the road for quite a ways and we lost them going into the brush. They went down towards a ravine. We could see some of the tracks and then we lost them. We searched around again for more tracks but they just all seemed to lead down that way.

The face was sort of a human-type face. It was pointed like on the top, it had a large pointed forehead on the top. A human's is sort of rounded off, his wasn't. It was dome shaped. It wasn't really that it had a lot of hair on its face, except the side features, it had hair on that. And he had a nose. It resembled a human's nose except it was flatter and wider than a human's nose would be.

His mouth was somewhat like a human's except it was bare, but a larger mouth. And he had something like carnivorous teeth, like a dog's teeth, and his eyes were good sized, and like I said, they shone in the dark. Real bright, like whitish-yellow, very very glowing. He had a bull neck, didn't have an extremely long neck, in fact there wasn't hardly a neck there, it was a bull's.

His skin had a yellow tinge to it like it had been bleached in the sun, like leather would be, buckskin would be, sort of tanned like you get in a taxidermist's and stuff, that's what it looked like. His torso was

107

GENTLE GIANTS

very muscular. I didn't get a good look at his hands because we didn't have the light actually on his body that much, you know, when we were trying to blind him in the eyes is when we actually shone the light on him outside. And he was bent over, stooped over. He had a pretty long stride to him. His legs were long and muscular and his waist tapered. He did have actually a large waist. I didn't really get a good view of his back. He had brown hair, real good shape, wasn't ragged or nothing.

He was about nine feet. I'm six-two, and he stood a good three feet taller than I was. Weight I'd say a thousand two hundred, something like that.

The witness said he had hunted Kodiak

bears and this looked about the same size, but didn't resemble a bear, except for its fur. He thought about it being someone dressed up, but that wasn't possible because of the size and the way it ran, it was "lightning quick." He said that the dog they had with them "he screamed too" and the hair on its neck stood up. "He just went berserk."

The giant creature sometimes made a sound like a moaning dog. None of the four in the car got any sleep, they could see the thing once in a while and hear it walking up and down. It never peered into the car, except one time it looked in the back window. The thing looked angry, but they thought it couldn't actually be angry or it would have ripped the car to pieces.

That story carries quite an atmosphere of violence narrowly escaped, but there is

This is a life-size picture of a sasquatch track of medium dimensions. Photo taken by Rene Dahinden on Blue Creek Mountain, September, 1967.

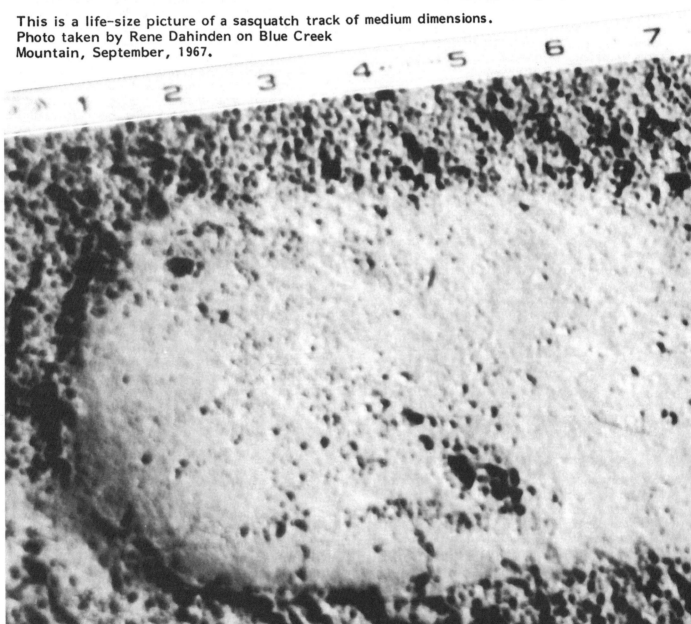

GENTLE GIANTS

no way of telling what would have happened had the young people not had a car to shelter in. Judging from the fate of people who had nowhere safe to run to, probably not much. An example would be the experience of Mark Meece and Marshall Cabe, from Darrington, Washington, near Cub Lake, 25 miles east into the mountains from Darrington, in August, 1969. The two boys were camping at the lake with six other teenagers and had hiked about a mile from the campsite. At about 8:30 p.m., still in daylight, they saw what the first thought

was a bear or a stump, but then it made a motion with an arm as if it were a human beckoning them towards it. Then it started to move and two others appeared from behind it. The boys started to run back to the camp and the three creatures chased them, running gracefully on two legs.

"At one time we looked back and they were 50 to 75 yards behind us," Mark Meece told a Seattle *Times* reporter. I'm sure they could have caught us if they wanted to."

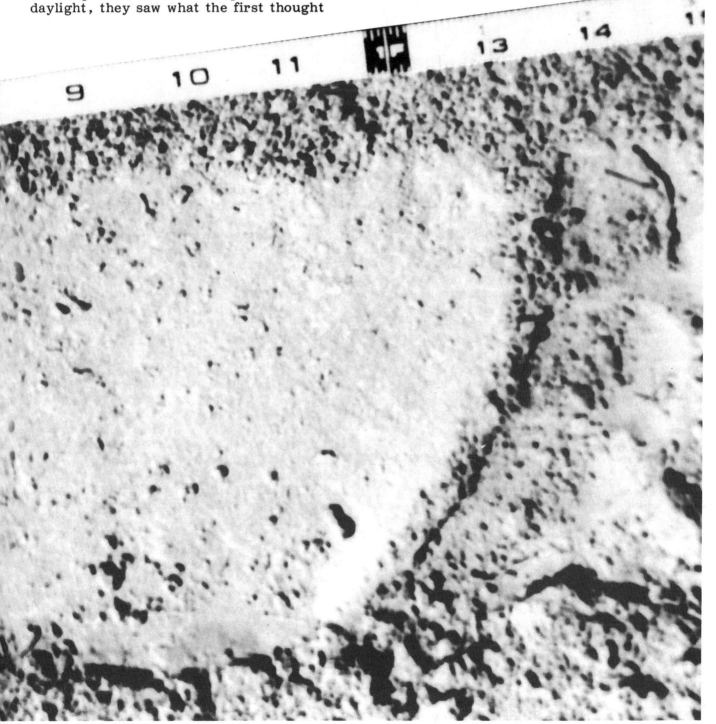

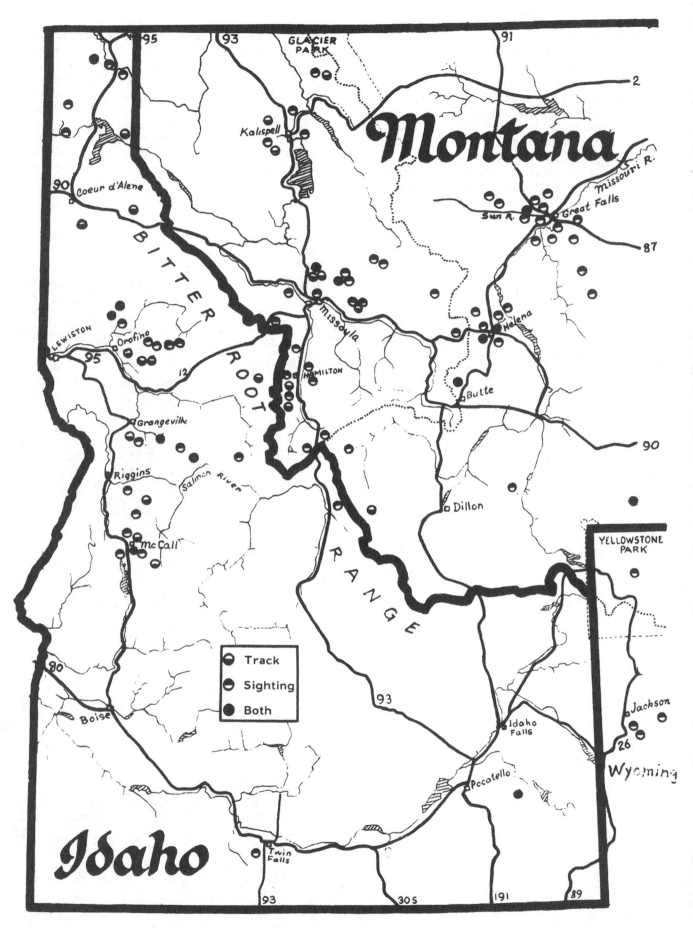

SASQUATCH SIGHTING AND TRACK REPORTS IN IDAHO AND MONTANA

Idaho

Sasquatch sightings have been reported across the full width of southern British Columbia and on the eastern slopes of the Rockies in Alberta, but south of the border the situation is different. Much of eastern Washington and nearly all of eastern Oregon is dry, open country; the exact opposite of the forested mountains from which most sasquatch reports originate. As would be expected there are very few reports from there. The same is true of southern Idaho, Utah and Nevada. In northern Idaho, however, there is an area that is both mountainous and moist, and there are plenty of reports from there.

The critical factor seems to be rainfall. Roughly speaking the reports are thickly grouped within the contours of a rainfall map indicating 20 inches of rain a year or more. Where less than 20 inches of rain falls in a year sasquatch reports are rare or unknown. The one major exception to that rule of thumb concerns the mountains of western Montana, where there are a lot of reports in areas on the dry side of the line, extending even to the plains in the vicinity of Great Falls.

It is rather puzzling when you study the map. There is a concentration of reports in northern Idaho, as would be expected, and a considerable spill-over into the drier areas to the east, yet from the equally mountainous areas of south central Idaho there are no reports at all—as if the rainfall rule applied in one direction but had been lifted in the other. I do not know either area at all well. Perhaps if I did some explanation would be obvious. Utah has three small areas shown with more than 20 inches of rain, and in one such pocket running along the east side of Great Salt Lake and Utah Lake, with an extension into the northwest part of the Uinta Mountains, all the recent Utah sighting reports are neatly fitted. Nevada, however, shows no area of 20-inch rainfall at all—the only state so universally dry—so all of its few sasquatch reports are exceptions to the rule.

The story written by Theodore Roosevelt about a trapper being killed by a sasquatch took place in the mid-1800's somewhere along the mountains that are now the border between Idaho and Montana, and from each of those states there was one newspaper story around the turn of the century.

The next report I have from Idaho dates from the 1930's. A woman wrote to me that members of her former husband's family told of returning to their ranch one day to find an animal kneeling on a bench and looking into a window. At their approach it got down from the bench and walked away upright. That was near Priest Lake, at the northern tip of the state, and she said that when she lived there herself in later years she often heard screams that were different from cougar screams.

Other than that I know of no reports from Idaho until the middle of the 1960's, but from then on there are many good solid ones. A lot of them have come to me from Russell Gebhart, of Lewiston, who has been an active investigator for quite a few years. He notes that there has been a whole series of reports from a wide area both north and south of the lower Salmon River in recent years, the earliest being in the fall of 1966 when two elk hunters found tracks in the snow abut 6,000 feet up on a mountain near Burgdorf. There were two sets of tracks, one estimated at 16 inches long, the other only eight to ten inches. They encountered them about a mile apart, but both were headed downhill and before they got down below the snowline they meandered across each other several times with no indication of joining up. One of the men, from Riggins, told Russ that after his own experience he made enquiries and found that there were other people in the area who had seen tracks but said nothing about them.

On June 27, 1968, the Grangeville *Idaho County Free Press* published a very interesting story written by Paul Kiepe. Here is part of what he had to say:

FRENCH CREEK, June 24—A search-and -photograph posse is being organized here this week to scout the backcountry 20 miles east of Riggins for further evidence of the two large, hairy, humanoid creatures who cut short the fishing excursion of a Council, Idaho logger at streamside on French Creek about 2 p.m. Sunday, June 16....

In all liklihood a representative of the Idaho Fish and Game Department will accompany the search party....

These developments have come about as a consequence of the vivid report of Frank Bond, 35, hooker for the Rice Logging Company of Council, whose fishing trip up French Creek "got shook up real bad" when an obviously female creature about seven feet tall and a companion creature at least a foot taller confronted Bond at a distance of not over 50 feet.

Bond, according to a telephone conversation I had with him last night, had just finished drinking at a spring about three miles upstream from the mouth of French Creek, and was turning back to fish, when the creatures rose up from behind a huge rock at streamside.

IDAHO

"I had a pretty string of fish," said Bond, "but what I did next was throw those fish at the creatures, turn around, and start running for all I was worth down the trail. In no time at all, it seems, I had covered the half mile to the car where my fishing partner was waiting for me, and I'm afraid I screamed at him: 'Let's get the hell out of here'."

Bond's description of the creatures resembles in many particulars the photographs and description of Sasquatches, hair-covered humanoids sometimes seen in California, Canada and elsewhere in the Northwest... except with respect to their hair color.

The reported Sasquatches thus far all have had black hair, whereas the all-over hair cover of the human-like giants seen by the Council logger was silver-gray.

"They were beautiful," says Bond. "I remember them that way even though I had only a short glimpse of them before I started running away. No one should harm them; try to shoot them, or anything like that," he continued. "They made no move as if to harm me. I was simply startled and frightened."

To the extent permitted by the circumstances, Bond's story is verified by Mick Burns, of Riggins, who was sitting in his car when the running Frank Bond, his fishing companion arrived.

"I thought he was going to run right through the car," says Burns.

After returning to Burns' home and digging out the story on the Patterson movie in *Argosy*, February, 1968, the two men and their wives returned to the scene of the encounter looking for evidence, and found two possible footprints 14 inches long.

The search to which the newspaper story referred was ineffectual, being conducted a quarter mile too far up the creek, as a result of faulty information. A week later, however, Wayne Twitchell, the owner of the ranch on which the encounter took place, was taken to the correct site by Mick Burns. A former professional trapper, Twitchell made a careful bush-by-bush search for hairs and found half a dozen of them, which reporter Kiepe later forwarded to Professor Ray Pinker, head of the Department of Criminology of California State College at Los Angeles. Professor Pinker had told the reporter on the phone that he could soon tell if the hairs were human or animal, but after microscopic examination he found that it had "definite characteristics of both animal and human hair." On August 22, in the *Free Press*, Kiepe quoted from Professor Pinker's report:

The hairs are not human hairs; we can state that fact positively because of certain unequivocal animal characteristics, but that isn't the whole story.

Exposing the French Creek hair sample in a comparison microscope beside each of the animal-hair samples that I have here has turned up nothing similar. Moreover, the scales of the hairs, the outside thatching seen microscopically, very much resemble the characteristic scales of the hair of human beings.

Professor Pinker went on to give the bad news about hair identification—without a sample of known sasquatch hair he had no way of being certain whether the unknown hairs were sasquatch hair or not. The collection of hairs he had for comparison purposes had been assembled with police investigations in mind and he was not even sure it covered all the known animals of Idaho. Later I was in touch with Professor Pinker myself and obtained for him samples of light-colored hair from all the animals native to that region, but he was still unable to identify the French Creek hairs.

In one of his later stories Paul Kiepe gave more details of the description given by Frank Bond. He said the smaller creature had large protruding breasts high on her chest, and because the large, hairless nipples were livid pink he thought she was probably nursing young. He also described a pungent odor somewhat like decaying flesh, or like the smell from a snake den, which remained in the area for a long time.

That same summer there was another incident farther north in Idaho, on the north fork of the Clearwater River, near Golden, in which something entered the camp of three prospectors, left tooth marks on a three-pound coffee can that indicated a tremendous spread of jaw, and made off with a five-pound sack of pancake flour which it apparently ate as it travelled, since they found the empty sack at the end of a trail of flour that went across a creek and up a hillside.

One of the men, C.E. Ricketts, from Pullman, Washington, wrote to me about it the following spring. He said it was most unlikely that the visitor was a bear, as it had ignored honey, jam, bacon grease and side bacon, and generally had not torn up the camp as a bear would have done. Also a bear should have left claw marks in some of the places where they followed the trail of flour. Searching around they found some tracks 12 to 14 inches in length and about five inches wide at the toes, but with very narrow heels. Then two of them came upon a cinnamon-colored animal which they did not

IDAHO

get a very good look at—they saw its back but not its head—which scrambled off hunched over in a very unbearlike fashion.

In 1969 there was a series of incidents at a sawmill near Orofino, east of Lewiston, Idaho. Russell Gebhart wrote about it to George Haas in January, 1971, and George published the letter in the *Bigfoot Bulletin*.

Mr. Moore is night watchman at O Mill, which is a small lumber mill about 20 miles from the nearest town. The mill operates the year round, but not at night. The entire area for miles around has been logged off and is quite brushy but there is some good timber left. The higher peaks are around 6,000 feet but most of the country is around 3,000 to 4,000 feet. There are quite a few abandoned mines around. Most are open pit mines but there are a few tunnels. I don't believe there are any active mines left. There are lots of bear and deer with a fairly good herd of elk in the area. The mill is located on a creek and has a logging railroad serving it. There is about three feet of snow there now.

Interview made January 16, 1971. In June, 1969, Mr. Moore, who lives near Orofino, Idaho, saw one female Sasquatch inside O Mill about 20 miles north of Orofino. The animal was described as being about six feet tall, completely covered with shiny dark hair except for its face, hands and the nipple area of its very large breasts. All visible skin was pink. He thinks it may have been nursing a baby. The eyes were large and fiery red. It was built like a very heavy, stocky person except that its arms were very long. It walked upright, just like a person, and moved gracefully. It appeared to have great speed when it wanted. It had a very disagreeable odor which he could not describe but would never forget or mistake for anything else. He observed the animal for about five minutes from around 35 or 40 feet.

He and several other employees had seen many tracks in and around O Mill all during the summer of 1969. There were at least three different sets of tracks, one very large set, one human size, and one child size. The tracks indicated that they had explored in, around and under the buildings. They played in the sawdust pile and ate sandwiches he had put out for them. They had apparently looked at themselves in a large mirror near the carriage. He could hear them jabbering among themselves and throwing boards which got in their way. They were in and around the mill most of the summer and he saw a very large, enor-mous dog once which he believes was running with them.

Reports continued to surface two or three times a year most years of the 1970's. One man from McCall reported two sightings; with his son and three other men he watched a black, heavy-set biped through field glasses from a distance of about half a mile, in the summer of 1970, then in the fall of 1972 he drove past a heavy-set, light brown creature on a forest road. In 1976 Russ got a letter from a man who had seen a sasquatch at Poet Creek campground, about 25 miles east of the Red River Ranger Station, the previous July. The man had slept in his car and woke about 4 a.m. He was headed for the restroom when he saw an animal walk out in front of him. It never looked his way but seemed to become aware of his presence and walked rapidly way with a very smooth gait. Russ wrote:

He described the animal as approximately eight feet tall and weighing around 800 pounds. It was covered with long, heavy, shaggy dark grey (not black) hair. Its arms were long and its legs slightly short. It had hands similar to human hands but he didn't notice exact features, although they were covered with short hair. Its neck was short and as wide as the head. Its face was flat and the forehead was very sloping. He could not see ears but there were lumpy-looking places where they should have been under the hair. It was built very massively but not fat. It appeared to be extremely strong. He detected no sounds or smells. He did not hear brush or twigs snapping but could hear the ground crunch under each step.

He described its posture as very peculiar and this feature seemed to catch his attention most. Its legs were always bent, as if in a slouch, and its buttocks stuck out far to the rear. Its back was in an extremely pronounced swayback position as it walked. When it came to attention upon sensing his presence its back became almost straight, but its legs remained in a slouched position. When it walked its arms swung similar to humans' and did not come to the ground.

Montana

For a long time I had no information at all from Montana and gave no thought to the possibility that there might be reports from there, so when a lady phoned in to a radio show in 1969 to say that she knew of such things in her childhoood in the early 1900's I didn't pay much attention. She mentioned the "Snowy Mountains" in Montana, and said that cattle were kept away from the mountains in the late fall and early spring because the hairy men who lived there would kill them to eat. Her father and her uncle had seen them on different occasions. I wish now that I had questioned her and followed up on the information. As it is I don't even know if she meant the "Big Snowy" mountains in central Montana or "Snowy Mountain" in the northwest.

Other than that and some very vague references to individual sightings about the same period, the Montana reports don't start until the 1950's. A publication called the *Bitterroot Journal* that used to be published at Hamilton, Montana, did a series of stories that included a sasquatch sighting near Seeley Lake in the fall of 1952 and a couple of other undated, but apparently early reports from the same area. Then in the fall of 1959, again near Seeley Lake, a man named R.W. Rye saw something very interesting, although he has not always expressed the same opinion as to what he thought it was.

He wrote to *Saga* magazine late in 1960 to describe the experience, which took place while he was bear hunting in the foothills of the Mission Mountains. He said that he was following large tracks in the snow and while crossing a small clearing got a feeling that he was not alone:

He looked around to his left and then to his right. There it was, twenty yards away, looking straight at him was this thing. Its head and arms were resting on a fallen tree that was five or six feet above the ground. Seeing only the head and arms, Rye thought it was a large bear, a very large bear. He began to raise his rifle to his shoulder. The thing, still leaning on the fallen tree, grinned at him—or so Rye thought. Then the thing let go an eerie, half-human scream. Its head began to rock from side to side and Rye could hear a rumbling sound. His rifle at ready, Rye back-pedaled. The thing moved a moment and Rye could see it from the waist up. It had a large flat head, stubby ears, a short neck and sloping shoulders with long arms. It was all covered with brownish grey hair.

That is how the story appeared in *Saga*. It resulted in the local news media getting in touch with Mr. Rye for confirmation, which he gave. The Billings, Montana, *Gazette* in a story December 4, 1960, noted that Rye had been a licenced game guide and had shot plenty of bears. The Missoula *Daily Missoulian*, December 2, had checked on another aspect of his story, about three men having gone missing in the same area, and found it to be true. The Missoula sheriff's office confirmed that a student from Montana State University had never returned from a trip to Crystal Lake; a Kalispell man had gone missing on a hunting trip to Swamp Creek, and a Missoula man who went hunting in the area was not exactly missing, but had been found killed, presumably by a grizzly bear.

Montana Sports Outdoors also took an interest in the story, and had this to say in its December, 1960, issue.

When we read the Saga account we immediately suspected the whole thing had been dreamed up by an imaginative editor shy a couple of pages of copy at press time. We were certain that Saga had picked the name "Rye" from a bottle tucked in a desk drawer and that Montanans were the victims of some "Rye" humor. Then someone checked the Billings phone book, and what do you know, the name "R.W. Rye" was listed ....
When we first asked if he could have been mistaken and had actually seen a bear, he quickly agreed. Too quickly, we thought. So we talked about hunting in the Seeley Lake area. Pretty soon we could see that here was a perfectly rational and experienced woodsman who knew the Seeley Lake area quite well, a man with an unpleasant experience that even he could not fully explain.
Then we again asked if he thought he had seen a bear.
"That was no bear," he replied firmly, "I've seen lots of bears and have killed a few, and I've never seen a bear that looked anything like that."
Rye thought the creature looked more like a huge ape ....

And what had Mr. Rye done after confronting this apparition? *Montana Sports Outdoors* told that story too:

At the time, Rye was armed with a .270 rifle and a shoulder-holstered .357 magnum pistol. What did he do next? Why he did what any other red-blooded, well-armed hunter would have done under the circumstances. He broke all records for cross country running getting the heck out of there.

114

MONTANA

One Montana story has a rather different twist in that it was the newspaper reporter himself who saw the animal—but he didn't write about it at the time. Tom Tiede was a reporter for the Kalispell, Montana, *Inter Lake* sometime in the 1960's. Some years after he left he wrote a story for the NEA syndicate in which he told of going to interview a hermit who lived in the woods near Columbia Falls. He failed to find the place, and was sitting down to rest before starting back when he heard something:

I swear the noise was like wailing, only not mournful wailing. Happy wailing? I can't help it, so help me, it sounded rhythmic, patterned—I shall say singsong.

I remember thinking it must be a hunter. And, knowing about hunters as I did, I remember wishing I had worn a red hat to distinguish me from a deer. I got up, to be as conspicuous as possible, and looked around. I had intentions of calling out, but thought better of it. And as the moments passed, and the singsong continued, my hours of viewing TV Westerns overcame me. I reached for my gun.

Suddenly, and I swear by the Abominable Snowman, I saw it. About 50 yards away. Coming down off one of the interconnected hills, passing at a moderate speed through the woods, disappearing and reappearing in the trees. I don't remember feeling anything. I could see plainly that it was not like anything I had ever seen before. It had swinging arms, like a B-grade gorilla movie, a gray coat of hair, and a small head which I could not make out. And it was moving parallel to me.

Now to be honest, I don't know if it saw me. But it stopped. And seemed to look in my direction. As it stopped, so did its song. I raised my rifle, forgot to take the safety off, but did nothing anyway. The thing paused for just a moment, then moved, silently now, off in a direction my shaky compass said was north....

I beat it out of the forest then, and to hell with the old hermit, I never went back. I never wrote the story before either.

That story was not published until 1971, with no indication as to when the events had taken place. My efforts to make contact with Mr. Tiede were not successful, but I was told by people at the paper in Kalispell that he had been there in the 1960's. No one seemed to be sure just when. A couple of Montana reports that do have definite dates came to me from Dr. Joseph Feathers, a professor at Western Montana College of Education at Dillon. He said that in the fall of 1960 a senior student named Dean Staton had seen huge footprints in the snow on a mountainside 15 miles southwest of Jackson in southwestern Montana. They looked like bare human tracks but were nine inches wide at the ball, five inches wide at the heel and 17 inches long. They went over a windfall four feet off the ground without breaking stride. After following the tracks about 300 yards across the open mountainside the student lost the trail on a rocky slope.

The other report involved two students, Gary Simons and Sid Richardson, who had taken Simons' Boy Scout troop from Butte on a campout about 15 miles from town up Brown's Gulch in May of 1964. At 4 a.m. one of the scouts was awakened by a form in front of his tent. Dr. Feathers wrote:

He got up and met face-to-face a Sasquatch man who was hairy all over, brown hair with silver tips. The man also had a heavy beard. The boy screamed and the man ran away. Gary came to investigate and they heard the creature splashing in the creek below. He made giggling sounds like a human.

Later in the morning they found many barefoot, human tracks, twenty inches long and six inches wide by measurement. The stride was seven feet. They could track him for one hundred and fifty yards before the trail also disappeared in rocky strata.

Years later another version of this report was published in a book titled, *Bigfoot: The Mysterious Monster,* by Robert and Frances Guenette. They had interviewed one of the scouts, who in the meantime had grown up to be "a serious man." Here is what Dan Maloughney, of Deer Lodge, Montana, is quoted as telling them:

I was with 11 other Boy Scouts and we were camping 18 miles south of Butte in Deer Lodge National Forest. We had walked into the forest 5 miles that day. We set up camp—numerous tents and lean-tos, built our fires. We ate dinner that night and just sat around and talked. About 11 o'clock we all went to bed and joked around until everybody went to sleep. Then about 2:30, 3:00 in the morning the kid next to me woke up, heard something in the woods, something going through our back-packs, knocking things over. Then all of a sudden this thing stepped on my friend. Then there was a lot of screaming, running around and hollering. We turned our flashlights on it, it was dark brown-black, it walked like a man, took great big strides, was three times my size, way over eight feet tall. We saw it moving away across the ridge for maybe four minutes.

115

MONTANA

There were other Montana reports as well, but not many. Then in 1974 I started to get letters from Greg Mastel, in Missoula. At first they seemed the sort of thing a young boy would write, but soon there began to be references to sasquatch reports I had never heard of. It turned out that Greg was indeed a young boy, at that time he would have been 10 years old, but that didn't stop him from becoming a highly-productive researcher. The information that he passed on to me was usually much briefer than I would have liked, but in the four or five years that he was active he uncovered far more Montana information than has come from any other source. The *Bitterroot Journal* also seemed to have no trouble in turning up new stories, most of them from hunters and loggers in the mountains west of the town. One of the most interesting involved a man named Ted Foster, from Dillon, Montana, who was cutting lodge pole with a chain saw just east of Lost Trail Pass in September, 1969, when he sensed that something was walking up behind him. Shutting off the saw he turned around and saw about 25 feet away a seven-foot, 300-pound "apeman" covered with black hair. It was just standing there looking at him.

Thinking to defend himself, Ted started the saw. The creature didn't move, but the hair stood up on its neck. Ted turned the saw off, the hair on the animal's neck settled back down again. Without taking his eyes off of Ted, the beast turned sideways and began walking away with huge graceful steps....The incident occurred about 3 p.m. in broad daylight.

The most interesting development in Montana in recent years has been the involvement of the Cascade County sheriff's department in the sasquatch investigation. This took place primarily in late 1975 and early 1976. Primarily the department was involved in trying to find out what was responsible for mutilating cattle in the county, but since they had nothing at all to go on they began looking into the reports of mysterious creatures as well.

The first report that the officers investigated was received on December 26, 1975, from two badly frightened teen-aged girls at Vaughn, a few miles west of Great Falls. They said that in the late afternoon they noticed some horses acting very strangely, pawing the ground and rearing, so went out to see what was the matter. They then saw a strange creature about 200 yards away which they estimated to be over seven feet tall and twice as wide as a man. One

Captain Keith Wolverton with his map of sasquatch sighting reports.

girl looked at it through a telescopic sight on a .22 rifle. She described its face as "dark and awful looking and not like a human's." Trying to frighten it, she fired in the air once, but it did nothing. When she fired again it dropped to the ground and pulled itself along with its arms a short distance before getting up on two legs again. The girls ran away, but one of them said she looked back and saw three or four of the creatures apparently helping the first one into a nearby thicket. A most unlikely story, but they agreed to take a lie-detector test, and both passed.

Next report came from Ulm, a few miles southwest of Great Falls, on February 21. Two boys reported seeing a hairy creature just south of the bridge over the Missouri River. One saw only a hair-covered arm extending from some brush. The other, who was ahead, described a very tall creature covered with dark brown hair and with glowing whitish-yellow eyes. Both boys passed lie-detector tests, but the story was not reported until two weeks after it happened so investigation at the site was fruitless.

In the meantime the sheriff's officers had talked to a man who claimed to have seen a nine-foot, hairy creature close up and two similar creatures farther away a couple of miles east of Ulm on the morning of February 22. He had stopped his car and run towards the nearest creature, which was walking in a field about a quarter mile from the highway, but when it turned in his direction he ran back to his car again. He also reported a grey object hovering in the sky

116

MONTANA

in the distance. Another man on the same day phoned in a report of a large hairy creature seen on a back road 15 or 20 miles east of Sun River that morning. He gave a name but could not be located later.

Shortly after midnight on March 7 a woman phoned the sheriff's department to report seeing a reddish-brown, hairy creature standing in the ditch, one arm forward as if it were about to cross the road, on the highway north of Vaughn. She refused to identify herself. Later that month a boy reported that he saw a hairy creature standing in the middle of Dempsey Road, in Great Falls, as he was riding his bicycle about 9 p.m. It ran off through a hedge into the yard of a house. That boy also passed a polygraph test.

I was in Great Falls early in April and was given a tour of the sighting locations by Captain Keith Wolverton of the sheriff's department. It was certainly not a proper situation for sasquatch sightings. Great Falls sits on the open prairie, with mountains on three sides but as much as 50 miles away. However all but one of the sightings was very close to either the Sun River or the Missouri, both of which have brushy valleys cut below the level of the plain. There were two more sighting reports in July of 1976 in the Great Falls area, and three more in August, then a lull until the following February. In March two young men reported that they had seen a "bigfoot" in a farm field and had set out to chase it with a four-wheel-drive pickup. At first it started to run away, but then it turned and faced them—and they turned and drove away. There was another sighting in May, just a brown, hair-covered biped seen walking on a sidehill, and then on August 30 Staff Sergeant Fred Wilson from Malstrom Air Force Base came to the sheriff's office with a most unusual story. Here is how it was recorded by Captain Wolverton:

He stated that on Saturday morning, August 20th, about 2:00 a.m., he and four others were camping near Belt Creek. There were two adults and three children—all males. They had gone in Friday. The weather was becoming stormy and rainy, so they decided to break camp and return to the car, which was about ½ mile away. Fred indicated that as they were returning they kept hearing sounds of twigs breaking. The three men thought it might be some sort of small animal. They were walking in the open pasture, and the sounds came from a bushy timbered area. When they reached the fence, which was near the car, one of the men looked back and said, "Fred, shine your
flashlight back there."

As the beam of light hit the trees, 74 yards away, they observed a large hairy creature standing next to a clump of bushes. The two young boys had already returned to the car ahead of the three men so they didn't see anything. When I asked Fred to describe the creature he said he was pretty shook up but remembered it had long tan hair, no neck, and stood on two legs. It seemed very muscular, the eyes glowed with the reflection of the light, and he stood 15½ feet tall. The three men then ran back to the car and one of them had a shotgun and he fired two shots in the direction of the creature, only he was shooting quite high. The shotgun was a 12 ga., with skeet shot. The creature started running toward them. Fred said it was another 10 feet to their cars, so they got into their cars and drove away. Fred had put his car in reverse. As he looked back he saw a portion of the creature and it appeard that he was just walking away.

Fred Wilson took and passed a lie-detector test. Next day he brought his two companions to the sheriff's office, but they asked to remain anonymous. All three said the creature stood and ran upright. They felt sure that if it were not for the two shots it would not have charged them.

A Tim Kirk cartoon from George Haas' Bigfoot Bulletin.

Nevada and Utah

There is a story to the effect that you don't require a talking dog to be eloquent, it's enough that he can talk. Similarly, it ought to be impressive enough that there are sasquatch reports from such unlikely places as Nevada and Utah, without expecting anything of special interest. Oddly, it doesn't work out that way. Among the few reports from those states are some that would be worth attention no matter where they came from.

Nevada, to begin at the beginning, has the oldest newspaper report of any western state. Dated August 4, 1870, it only wins out by six weeks, and it's a highly unbelievable yarn too boot, but nevertheless it is the first. Here it is, as printed in the Lansing, Michigan, *Republican,* but blamed by them on the New York *Tribune:*

The Latest "Wild Man"

The wild man has now turned up as far away as Nevada. Like the sea serpent, he appears from time to time, in different parts of the continent; and the stories which the country papers get up about him are always very wonderful and awful.

The last appearance of the wild man or rather "the object" as he, or she, or it is called, was in a desolate region of Northern Nevada, where an intense state of excitement has been roused about it. A large party, armed and equipped, lately started in pursuit of "it" and one night a splendid view was obtained of the object which, it was concluded, had once been a white man, but was now covered with a coat of fine, long hair, carried a huge club in the right hand, and in the left a rabbit. The moment it caught sight of the party, as the moon came out, it dashed past the camp "with a scream like the roar of a lion," brandished the huge club and attacked the horses in a perfect frenzy of madness.

The savage bloodhounds which the party had brought along refused to pursue the object; and so the party hastily raised a log rampart for self-defense; but, instead of making attack, the object merely uttered the most terrible cries through the night, and in the morning disappeared. It was evident, however, from the footprints, that the object would require a "pair of No. 9 shoes," and this is all we know. The party could have shot it on first seeing it, but failed to do so.

From a careful reading of this account, we should judge that the wild man or object made tracks immediately after seeing this party, for Salt Lake, to keep company with the sea-serpent which the Mormons who lately saw it there say is a half a mile long, and looks dreadfully terrible.—N.Y. Tribune.

A purist might argue that it isn't really the oldest story, because the one printed in September, 1870, in the Antioch, California *Ledger* referred to a sighting that had taken place the previous fall. On the other hand this story doesn't indicate how long ago the events had taken place. As to the use of a club and the attack on the horses, however, there really is no defence to be offered. Behavior like that is too great a departure from anything observed elsewhere to be acceptable from any source, let alone from a story presented as this one is. There remains a possibility, however, that it is a story embellished in the telling rather than one drawn entirely from someone's imagination.

The next Nevada report, just over 90 years later, is a good solid one. Dean Pollard, of Myrtle Creek, Oregon, while engaged in counting deer in the Diamond Mountains north of Eureka, Nevada, in November, 1960, observed tracks in the snow on two successive days. They were about 15 inches long by seven inches wide, showed five toes, and sank 14 inches deep where his own tracks sank six inches. The altitude was about 7,000 feet, and the tracks were headed downhill.

The next specific report I have—skipping a couple of vague ones—is from close to the California-Nevada line, and was contained in a letter to the editor of the Reno *Evening Gazette,* published August 11, 1973:

Bigfoot sighted?

EDITOR, the Gazette: We recently visited Lake Tahoe for the first time (a beautiful place) and, while on our way up Kingsbury Grade we had a very frightening experience.

It was about 8:30 p.m., July 29, about two-thirds of the way up the grade, which you know is very narrow and steep. As we came around a turn, we saw something on the side of the road which we thought was a black bear. As we got closer, it was standing on its hind feet in an upright position.

When it saw our car, it went into the bushes. Just before it disappeared, it turned and looked at us.

Its face was more flat, like a gorilla's. It was about seven feet tall, and very shiny.

There was no place to get off the road, so we went on to Tahoe.

There were four people in the car and we all witnessed the same thing.

When we first saw it, my friend began yelling "Blackfoot! Blackfoot! I mean, Big-

NEVADA and UTAH

foot!" We don't know for sure what kind of animal it was, but we do know we saw it.

We went to the sheriff's office at Lake Tahoe and told our experience. They said they had had another report that two girls had seen the same thing and were still crying with fright. The excuse they gave us was that it was a crazy person in a gorilla suit trying to scare people.

We don't buy this story, after seeing it.

I am a sportsman. I have hunted deer, buffalo, antelope, elk, bear and lion, and I know an animal when I see one. This definitely was an animal.

Until they catch that so-called imposter, we still have our own ideas. Did we see Bigfoot?

We are sure some of you will say we are crazy or that we had too much to drink (not one drink). But maybe there is someone else who has seen this animal?

Mr. and Mrs. D. Cowdell
Mr. and Mrs. C. Searls
Murray, Utah

If anyone else had seen such an animal they didn't stick their necks out by writing to the newspaper.

In July, 1979, there was a newspaper reference to sighting of a 7-foot, apelike animal on the Pyramid Lake Indian Reservation north of Reno, but I know nothing specific about it. In January, 1980, however, there was a solid-sounding report from the nuclear testing site in southern Nevada. An employee of Reynolds Electrical and Enginerring Co., a contractor on the site, reported seeing the creature on January 16, about noon, while driving along Tippipah Highway at the northern end of the huge test area. He said that it was between six and seven feet tall, walked erect like a man, and was completely covered with dark hair. He stopped his truck about 40 yards from it and watched it cross the highway and disappear into the sagebrush. A spokesman for the Department of Energy described the man as "reliable" but did not identify him.

From Utah I have no early reports on file, in fact until 1977 I knew of no reports from there at all. There were two sightings that summer and there have been three more since, that I know of. In addition the Bigfoot Research Society, of Dallas, Texas, turned up a Utah report from about 1959.

The 1959 report is one of the few that has been put in sworn form, and it is interesting also because the witness, John Paul Ingram, of Mountain Home, Texas, is now a professional guide and trapper. He is not certain of the year, but the month was October, the day before deer season opened. He and another hunter were out scouting the area near Monticello, in southeastern Utah, in an open jeep. Going up a hill on an old mining road they rounded a curve and saw an animal come from left to right across the road. It was about four and half feet tall, estimated about 75 pounds, and walked upright bent slightly forward. Covered with short hair of a smoky black color, except on the face, ears and hands, its skin was almost as dark as the hair. The face and ears were humanlike. They were about 40 feet from it when it disappeared in knee-high sagebrush on the lower side of the road, and it apparently hid, as when they got out of the car to look down the hill they could not see it. The other man insisted on driving on "before its mother comes along."

The 1977 sightings came to light in reverse order, with a clear indication that the earlier one would never have been reported except for the publicity given to the later one. On August 25 the Ogden *Standard Examiner* carried the story told by Jay Barker and Larry Beeson, both of Ogden, who, along with six youngsters, had seen a strange creature in the High Uintas mountains on August 22. The two family groups had met while hiking, on a ridge about 12,000 feet in elevation above Pass Lake. Looking down into Cuberant Basin they noticed an animal standing beside a small alpine lake below them. It was light colored about the head and shoulders, dark below, and Mr. Barker thought at first it might be an elk, but when the boys knocked loose some rocks which rolled down from the ridge the animal looked up at them and then walked off on its hind legs. They had it in view for about four minutes as it moved through scattered trees before disappearing into heavy timber. It was a long way off, but it looked to them about 10 feet tall.

That wasn't a particularly special report except for the number of witnesses, but the story told by the people who came forward after hearing about it had a very unusual story indeed. Two couples, Mr. and Mrs. Robert Melka from Bountiful, Utah, and Sergeant and Mrs. Fred Rosenburg from Hill Air Force Base, had sat on a ridge in the Uintas and watched three such animals, two of which were "romping" in a mountain meadow. Here is part of the story as told in the Bountiful, Utah, *Davis County Clipper* September 2, 1977:

"We sat on a ridge looking into a meadow only 300 to 500 yards away when we saw the first creature," said Mr. Melka. "A few seconds later a second beast—both much larger than humans—entered the meadow and the two romped back and forth in the clearing for at least ten minutes."

He said the weather was clear and there was no obstruction.

All four spoke very emphatically about what they had seen, and the four, interviewed separately as couples, reported seeing exactly the same things and described in detail what the creatures looked like and their actions.

"These things, whatever they were, walked and ran only on two legs. They had arms, legs and bodies much like a human being but covered with hair. Only their hands and feet did not have hair," they said.

"They were profile to us much of the time and there is no question about them having only two legs—but their legs and arms appeared out of proportion (larger) to those of humans.

"Two of them would romp and play in the meadow while a third stood some 100 yards away at the edge of the meadow near the point of a cluster of pines. The three were in perfect view all of the time except for a few seconds when two of them disappeared into a wash, but returned again to the meadow.

"They could not have been bears or other animals. We don't know what they were but they were something that we have never seen before. They did not have pointed snouts as bears do and we had a good, long look at their profiles."

Sergeant Rosenberg said he has hunted wild game for more than 25 years around the world and "these were nothing like anything I have seen before, in real life or otherwise."

The beasts were described as being 8 to 10 feet tall with much broader shoulders than humans but with distinct necks, arms, hands, legs and bodies much like humans.

"It's hard to judge our exact distance but I know that if they had been elk, I could have killed them with my high-powered rifle that has a telescopic sight. We were that close," said Sergeant Rosenberg.

They agreed that although they had the appearance of humans, they did not act like humans.....

The sighting was on July 10, but the four never discussed the matter further among themselves or with anyone. The Melkas never even told their children and the Rosenbergs told their family only that they had seen something strange while in the Uintas.

"We thought a lot about it, but didn't want to talk about it and be humiliated by those who wouldn't believe us," said Mr. Melka.

Sergeant Rosenberg said that when the first sightings were reported last weekend by Jay Barker and his party from North Ogden, "we decided to tell the world what we had seen earlier in the month in the same general area of the primitive Uintas.

The sighting by the Davis County couples was about one-half mile southwest of Elizabeth Lake, about 17 miles northwest of where the Barker group saw their "beast."

"It's a very primitive area, about 18 miles from the nearest road," said Sergeant Rosenberg. We had taken a four-wheel drive jeep into the area looking for a suitable site for an elk camp in late September. We stopped and walked to the ridge to look into the lake area.

Pat (Mrs. Rosenberg) was the first to catch the creature out of the corner of her eye. At first we were shocked at the huge size of these things but then we became intrigued and just stood and watched them. We weren't scared, but we were apprehensive.

There have been three reports of sasquatch sightings in Utah in 1980, all fairly mundane ones. Pauline Markham, a resident of South Weber, in Davis County, said she saw a dark creature, appearing to be eight or ten feet tall, walk along a ridge behind her house in daylight on February 3. The following night Ron J. Smith, who also lives in South Weber but two miles away on the other side of town, saw a seven to eight foot figure in his pasture when he went to feed his horse. He had the impression it might be a husky person in a fur coat, but after it walked into a clump of trees he heard four loud, inhuman screams.

On February 25, Lee Padilla, of Clearfield, was driving in Riverdale at 3:30 a.m. when a 10 to 11-foot animal loped across the highway about 25 feet away. He said it had long legs, a head like a gorilla, and long, dark brown furry hair. It seemed to be both very fast and very graceful as it ran across the road, and he guessed its weight at 600 pounds.

The Prairie Provinces

In the early years of the sasquatch investigation the idea that there could be anything going on east of the Rockies would not have crossed our minds, and for a dozen years my activities were limited to British Columbia and the three states immediately to the south. There were the tracks seen by David Thompson in 1811 near the present site of Jasper, but there was no way of knowing if they were connected with my subject or not, and otherwise there was no definite information at all except one story from far-away Labrador. When a break came in that situation it wasn't in Alberta, which shares the Rocky Mountains with B.C., but off in the center of the continent, in Manitoba. And although in the past few years we have learned of a great deal going on in Alberta, when it comes to stories with something special to them Manitoba still holds the edge.

The first surprise came in a clipping from the Winnipeg *Free Press* in the summer of 1968. I don't remember now where I got it, and it doesn't have a date. It told of the president and general manager of the co-op at a place called Easterville coming to the *Free Press* office to tell them that on two occasions people driving between Easterville and Grand Rapids had seen a hair-covered monster "somewhere between a moose, a bear and a man," on the road. The last line of the story said the creature had a foot like a moose, and that was quite enough to kill any interest I might have had— but I learned better later on.

About that same time something interesting was going on in the uninhabited upper valley of the North Saskatchewan River in Alberta. At a place called Kootenai Plains a group of Indians from a reserve near Edmonton had set up a new home for themselves in tents and teepees, determined to make a new life for themselves and their families more like that of their ancestors and less influenced by the undesirable elements of the white man's culture—particularly alcohol—than they found possible in less isolated surroundings. Before long they found themselves encountering something from the old days that they hadn't counted on—the hairy giants of tribal legend.

The creatures didn't cause them any trouble, and no one made much fuss about them, even though big tracks were seen within a couple of hundred yards of camp. One man who saw one watching him at close range as he was cutting brush told an Edmonton *Journal* reporter, "I didn't know what to do, so I just went on chopping wood." After a while the creature went away. Some photos of 17 and 13-inch tracks were taken that year, but it wasn't until 1969 that the real fuss started, and it did not start with the Indians.

About 20 miles downstream from Kootenai Plains a dam has been built that now floods all the bottomlands. In 1969 work on the dam was just in its preliminary stages, and down by the river, on August 23, a crew of five men was at work installing a pump. Harley Peterson, from Condor, Alberta, was the first to notice that they had an audience. At the top of the steep riverbank, about half a mile away was a dark upright figure that looked like a large man. He kept glancing over at it from time to time, and the figure remained upright for about half an hour, then it sat down for about 10 minutes, then stood again for another 15, and finally walked along the ridge of the bank until it disappeared in some trees. During this time the number of men watching the creature had increased to five. The thing was on the same side of the river as they were, but across a deep bend which put two widths of the river plus a sandbar between them, so they could not readily get closer. They tried waving, but the figure did not react.

They had neither a camera nor binoculars. One man went to the main camp to get a transit from one of the engineers, but got into a discussion about why he wanted it and finally rushed back without it for fear the thing would disappear while he was gone.

None of them realized the true size of the thing they were looking at until two of the men went over to the place where it had been while the others watched from below. As the figure had walked off they had carefully noted its height in relation to the small trees in the background, and they were astonished to see that the men in the same place appeared to be only about one third as high. At the place where the figure had sat down it had rested its back against a bump on the ridge, with its head still showing on the skyline. Now they discovered that a man in the same position was no higher than the bump when he was standing up. After considering their comparisons they decided that the thing must have been at least 12 feet high, probably 15 feet.

Their story was widely publicized, and I went to the site about a week later. The crew who had seen the thing were employed by a subcontractor whose job was finished, so none of them were on hand to show me around, but I looked over the area and then went on to find some of the witnesses. As far as I can recall I questioned two in person

THE PRAIRIE PROVINCES

and one on the phone. All told basically the same story. Most of the ground at the site was hard packed. There were a few patches of sand that would have showed footprints but no one had found any. By the time I was there everything had been trampled many times and there were even planes flying overhead looking for the creature.

What interested me most was the fact that there was a trail right at the edge of the cutbank and that beyond it the land sloped down away from the river. That meant that the figure must have been walking right at the edge of the bank, otherwise the feet and legs would have been cut off from view. The men who went over there would also have walked on the trail, so they and the original figure would have been exactly the same distance from the viewers and from the background trees against which their heights were compared. There would be almost no margin for error in the method the men used to arrive at their spectacular estimate of height.

Later I watched from the pump as two other men walked the edge of the bank, and I found that their appearance conformed exactly with what I had been told. They were indeed just dark figures with arms and legs that could be seen moving, but it was to far away to tell that they were clothed, and they were less than half as high as some of the trees the giant was said to have overtopped.

Of course I didn't see any giant, and I have no proof that the five men didn't make him up. But I do know that if they did see him and did compare his height to the individual trees they had pointed out in my poloroid pictures, then he was something well over twice the height of a man.

That is not the only story about such a giant, of course, but eight or nine feet is a much more believable height than 12 or 15, and it would be very comfortable if the outsize estimates could be dismissed as just the result of too much excitement. That can't be done with this report or with the one of the two prospectors who also scaled the thing they were watching against the trees, but at much closer range.

"Tell me more about these 'little, hairless, cloth-covered creatures' you've been seeing . . ."

THE PRAIRIE PROVINCES

I have encountered references to a suppo-sedly well-known tendency people have to overestimate the size of things they see out of doors, but I wonder if there really is such a tendency. My own impression is that most people are fairly experienced at estimating the height of erect bipeds within anything like human range, and the sug-gestion that they would tend to turn an extra two feet into four or six is not too convincing. I am bothered also by a recol-lection of my own. At a place where our family often went for picnics there used to be an oar leaning against a tree, about 30 feet off in the bush to one side of a trail. I saw it many times and it never looked like anything but an ordinary oar. Then one day I needed an oar and since that one was obvi-ously abandoned I ploughed through the underbrush and got it. To my amazement it proved to be 10 feet long, and altogether the biggest oar I ever had anything to do with. Leaning against a tree with part of its length hidden by bushes—much like the situation in which many sasquatches have been seen—it did not look its size at all.

I also reflect that Christmas trees cut in the woods (which is something I am old enough to remember) are invariably far too tall to put up in the house until some of the butt is cut off.

Rene Dahinden and I were very hopeful that around Kootenai Plains we had found a better place to look for sasquatches, because the tree cover there was thin enough so that animals could be tracked from the air if there was snow on the ground. Nothing came of that, however. The Indians are still in the area and may still be encountering the creatures, but if they are they wouldn't say so because they don't want them dis-turbed. The only substantial report I know of from that area since that time was carried by Canadian Press in August, 1974, three months after the event described. It said that Ron Gummell, of Calgary, on May 11 1974, had driven round a bend on the David Thompson Highway beside the lake that now covers the area west of the dam, and was confronted by two creatures standing in the middle of the blacktop. He estimated their height at 12 feet, and he was in a good position to do so, having stopped only about 30 feet from them. They stared at him for about 10 seconds and then jogged away.

"They were so big they could have picked up my car and thrown it in the lake," the story quotes him as saying.

There still are not a great many reports from the Rockies, from either the Alberta or the B.C. side, but there is no one in that part of the country working to dig them out. I have learned of two because the witnesses wrote to me, and that would indi-cate that there are probably lots of others. In one case three amateur prospectors had a close look at a seven-foot female that sat on its haunches and watched them eat lunch not far from Banff one day in September, 1969. In the other a hunter saw a black ani-mal that looked like an ape standing in a river some 20 or 30 miles southwest of Caroline. He ran towards it, but it waded to the far bank before he got very close. It was black in color, eight or nine feet tall, and when first seen was splashing itself in the river. That was in September, 1976.

By no means all of the Alberta reports have come from the mountains, or even the foothills. In the fall of 1972 and again in October and December, 1973, there were sightings involving four persons at various points along Seven Persons Creek, south-west of Medicine Hat. Fifteen-inch tracks were found in the ice on the creek near the location of one of the sightings, and there were other tracks found that were not con-nected with any of the sighting reports.

The October, 1973, report was unusual in that the witness never identified himself or herself, but sent two notes to a rancher, Leonard Edvardson, who was investigating the other reports. With the notes, which were signed, "An Observer," were two sketches of what "Observer" had seen, and they indicated clearly that it had appeared to be a female carrying a baby. Whatever it was, it was six or seven feet tall, appeared to be covered in dark reddish-brown fur,

What "Observer" saw.

and had been observed for about four minutes, in twilight, at a distance of about 250 feet. "Observer" decided, until seeing the other reports, that it must have been a Hallowe'en prank, although it was puzzling that it was carried to the point of wading off in the cold creek. While it is a long way from the Rockies, Seven Persons Creek flows out of the Cypress Hills, which rise to nearly 5,000 feet.

The sighting of a large female and two small sasquatches near Sexsmith by veteran trapper Bob Moody in October, 1975, was also many miles from the mountains. Game guide Art German described the area as being just where a jungle of poplar and willow merges with an expanse of pine timber, with a river nearby and behind that a hill and then a swamp, all completely wild country.

East of British Columbia and the Yukon, Canada is not mountainous, but most of it is covered with trees, including much of the northern part of the Prairie Provinces. Until 1968 this vast northern forest had not produced, to my knowledge, a single specific sasquatch sighting report. Since then there have been about a dozen from Manitoba, a few from Alberta, and finally, in 1978, a report from northern Saskatchewan.

The encounter had taken place in July of 1972, but the witness, Kamil Pecher, was a recent immigrant from Czechoslovakia, had no idea what it could be that he had seen, and soon found out that no one would believe that he had seen it. He kept the story pretty much to himself until he read in *Fish and Game Sportsman* magazine a brief report about suspected sasquatch footprints near Theodore, Saskatchewan, in 1977. By then he was a Canadian citizen, had been paying attention to sasquatch reports from elsewhere, and was sufficiently sure of his ground to write an article for the same magazine about his own experience. (Later he went into the matter further, discovering that some of the "Wetiko" stories of the Cree Indians did not seem to be about supernatural beings but were more like sasquatch reports, and he wrote an article about that which was printed after his death in the Fall, 1979 edition of *Pursuit,* the publication of Ivan Sanderson's Society for the Investigation of the Unexplained.)

Mr. Pecher made a solo kayak trip on the lower Churchill River in the summer of 1972, planning to travel from Missinipe, Saskatchewan to Pukatawagan, Manitoba. While portaging his gear around Grand Rapids he was disturbed by the loud breaking of branches in the thick bush near the trail,

so much so that he dropped his pack and started looking around. Walking back, in the trail he had already traversed, he found a track that had not been there a few minutes before; a track four times the length of his clenched fist and twice as wide–40 centimeters by 20 centimeters (16 inches by eight inches). The next track was three of his own paces away. There were no claw marks.

After carrying everything to the end of the portage he went back with his camera to photograph the tracks, but found that in the deep shade under the trees there was not sufficient light. He started looking around for tracks that might be better lighted, but was halted by a strange smell and something in his peripheral vision. Looking around he saw a huge dark shape, with wide shoulders above the level of his head and an appearance like a man or an ape. He fled to his kayak and out into the lake, so unnerved that he no longer felt safe to finish his trip and instead returned home.

It does not appear from his articles that Mr. Pecher was ever aware of the events in Manitoba the following year that not only supported his claim that there were sasquatches in the northern forests, but also identified them with the Wetiko of Cree legend.

In January, 1974, Brian McAnulty, a police reporter for the Winnipeg *Free Press,* did some checking around after someone had shot a cougar—a rare occurence in Manitoba —and stumbled on something much more unusual. Just a couple of months before, the Manitoba Museum of Man had received a remarkable letter from a wildlife conservation officer at The Pas, which is well north of the prairie country in the timbered, lake-strewn Canadian Shield. The subject of the letter was a set of footprints 21 inches long, and the witness was the conservation officer himself. The letter went as follows:

On September 29, I found some human footprints which I believe you might be interested in.

These tracks were near Landry Lake, 20 miles west of The Pas in Tsp. 56, Rge. 23 WPM. The enclosed photo shows the size of the print in relation to an 18" ruler. The depression was approximately 1¼ inches deep. My footprints in the same soil sank less than ½ inch. There were imprints of four large round toes almost in a straight line across the front. There was a slight depression in the ground before the print was made so that the two toe marks, across from the ruler in the photo, do not show up well. The print is generally flat from heel to toes. *Distance between prints was approximately 20 inches.*

The area in which I found the prints is a limestone base ridge covered with spruce and jackpine surrounding the north and east side of Landry Lake. The prints, three of them, were in an old ant hill that had been spread. Aside from this one soft spot, the ground cover in the area is mainly light moss and vegetation over a base of clay-gravel mixture and does not lend itself to tracking animals.

I was moose hunting at the time I found the three tracks. My impression at the time was that they were not man made and that they fitted no animals in the area. They were sharp and clear at the time and were made after the rainfall on September 25. I did not carefully compare all three prints at the time but they appeared to be identical except for the toe marks.

On October 1, I returned to the area with R.J. Robertson, Wildlife Biologist, and B.E. Jahn, Wildlife Technician, to look at the prints again. That morning a moose hunter had walked through the area and stepped

The 21-inch track found and photographed by Conservation Officer Bob Uchtmann near The Pas, Manitoba.

in two of the prints, making it impossible to compare the three. A light rain was falling and had washed the sides of the impression in so that plaster of paris casts were not practical. This, also, resulted in poor photography.

There is the possibility of someone having made the prints but this strikes me as being highly unlikely. The prints were approximately ¼ mile from a bush trail passable to vehicles. Back along the road 200 yards is a dike across a creek on the north end of Landry Lake. It seems more likely that someone creating a hoax would have placed the prints on the dike where there would be more chance of seeing them. As for the other hunter on Monday walking through the area, I believe this to be pure coincidence. I, also, had placed a red ribbon by the tracks to which he may have been attracted.

The letter concluded with some points to note in studying the photos, and was signed *R.H. Uchtmann, Conservation Officer.*

Nothing ever came of the tracks at The Pas, but because of them Brian McAnulty decided to check whether anything more had happened on the road between Easterville and Grand Rapids (not the same Grand Rapids as the scene of Mr. Pecher's encounter. One is on the Churchill River in Saskatchewan and is just a rapids. The other is on the Saskatchewan River in Manitoba, and is a town). It turned out that things had never stopped happening. Not only had there been further sighting reports by Indian residents of Easterville, there had been three incidents in which the witnesses included non-Indian schoolteachers from Easterville. All of the sightings had been while driving the road at night, but the witnesses told Brian there was no doubt about what they had seen. A woman who had taught at Easterville school for five years saw something twice, in the summer of 1969 and in November, 1970. On the first occasion an Indian couple was with her, and all of them saw a dark manlike figure run from the road, hurdling over willows and small bushes with long strides. The teacher was no longer at Easterville when I went there to get computer questionnaires filled in in 1974, but one of the Indians told me the thing was the color and size of a moose; yet jumped over things like a man.

On the second occasion a dark, heavy, shaggy form beside the road which they at first thought was a man turned out to be seven or eight feet tall and covered with hair. It had an extremely short neck, was leaning forward a little, and had long, hanging arms. They backed up to get a better look but it had vanished in the trees. The

THE PRAIRIE PROVINCES

other witnesses that time were the principal of the school and an Indian who had died before Brian learned of the story.

The third sighting, in April, 1973, involved yet another teacher, in fact there were two of them in the car but one had dozed off. The driver had to slam on the brakes, going into an uncontrolled skid, in an attempt to avoid a dark thing that was walking on the road in the same direction he was driving. He estimated it to be nine feet high and very broad, and as he braked he saw it turn sideways, giving him a look at a flat-profiled face.

All of those teachers had left Easterville by the end of 1973, but Brian got in touch with them and did a full-page feature story in the *Free Press* on January 26, 1974, about these events and about the creatures the Crees call "Weetekow" and the Salteaux call "Wendigo". He also told the story of a man and woman driving from Werner Lake, Ontario, towards Winnipeg, Manitoba, in December, 1972, who saw a tawny creature "sort of like an ape" run across the road in front of their car at night. They estimated its height at seven feet, and they had backed up and found tracks in the snow too far apart to be made by a man.

"It's like walking down a back alley and bumping into a Frankenstein monster," the young man told Brian. "Everybody knows there is no such thing, but you've just seen him."

By the time I met him in 1974 Brian had also found two people who had seen sasquatches near Reynolds, Manitoba, southeast of Winnipeg. Since then he has kept track of several more sightings in the Grand Rapids and Easterville area, and has invesigated several more in the area from Beausejour to the Ontario border northeast of Winnipeg. It should be noted that if Brian had not stumbled on the first footprint report while working on a story about something else none of this information might ever have come to light. I don't know if it is probable that much more information could be found in Saskatchewan if someone were to investigate, but considering the amount that has come from Alberta without anyone looking for it I suspect that a Brian McAnulty working there would be buried in stories by now.

There are also two good, recent reports from Manitoba that did not come through Brian. One was relayed to me by the Royal Canadian Mounted Police, at Vancouver, who had received the following telex from their Norway House detachment:

Norway House to Vancouver Lower Mainland Division, attention Sergeant Doane.

It was reported to our office on the 26 July 76 by the chief of the Poplar River Indian Band that many of his people have sighted on the reserve many times a large hairy animal that walks on two legs. Poplar River is located approx. 76 miles to the south of Norway House. An investigation was conducted and the results are as follows:

Several people were interviewed and they all stated that the animal was approximately seven to eight feet tall and was very broad at the shoulders. It had the general body structure of a man only many times larger. A foot cast was taken of the foot impression that was left behind by the so-called monster and is held at this detachment. It measures 16 inches by five inches, and has only three toes. Its fur is a glossy gray color and it has white hair on its head. They stated that it was very powerfully built and one man reported that he saw it swimming. To date there have been no further reports of sighting in our area. It should be noted that this so-called monster seemed very inquisitive towards the people and would come around the houses on the settlement and look in the doors and windows.

This is about all we have so perhaps you could pass this on. Thanks in advance.
Downing, Norway House detachment.

I turned that over to Brian, but Poplar River is not an easy place to get to—even Norway House is far from any highway—and I have heard no more about it.

The most recent report was in the fall of 1979, from the Little Saskatchewan Indian Reserve near Gypsumville. Newspaper accounts are long on "gee whiz" and short on specifics, but they seem to add up to at least a dozen witnesses and several sightings of a nine-foot humanlike creature covered with black hair. More impressive is the reaction of the local conservation officer, Ron Heroux, who made a cast of a 15" footprint found in hard mud. He estimated that whatever made the impression must have weighed at least 400 pounds. He said that a 250-pound man jumping up and down on the spot hardly made a mark.

Screams in the Night

Quite a few reports have accumulated over the years in which the witnesses describe sounds made by the sasquatch, but a description of a sound is little use. Other people have tape-recorded sounds assumed to be made by sasquatches, which may be of value but there is no way to be sure. Only one of these recordings is of good quality, and there is a fascinating story to go with it. I have talked to a couple of the people involved, who seemed to be entirely sincere, but have had no opportunity to investigate.

The story is told fully by Alan Berry in the paperback book *Bigfoot* which he co-authored with the late Barbara Ann Slate in 1976. He also markets a record of some of the sounds, which are a remarkable series of growls, grunts and gibberings apparently intended to intimidate rather than as communication. The sounds were first heard by Warren Johnson, of Modesto, California, and his brother Louis in August, 1971. Warren Johnson wrote to Ivan Sanderson about them the following year, and his story and Alan Berry's, although there are minor inconsistancies, are basically the same. The two brothers and some other men used a crude log shelter high in the Sierra Nevada, as short distance from Yosemite National Park, for a hunting camp. They have not made public the exact location but it is not far from Strawberry, which puts it in an area where there have been other sasquatch reports.

On the occasion when the sounds were first heard the brothers had cooked a meal on an outdoor stove near the shelter and had then retired inside. A short time later they heard a commotion among the cooking things, followed by a torrent of sound that thoroughly frightened them, with tooth-popping and chest beating apparently mixed in with the vocalizations. After everything was quiet they went out to look around and found enormous five-toed footprints, nearly 19 inches by nine, in a muddy patch where some tea had been spilled.

They heard the noises again on that trip and again a short while later when five of them stayed at the shelter. On that occasion Louis, peering through a hole in the shelter wall, saw a shape enter and leave a patch of moonlight. Johnson's account gives its size as 10 feet tall and four feet across the shoulders. These experiences continued, and during the same period a variety of tracks were found in the vicinity, the largest 22 inches long, the smallest seven inches. They made casts of some of the

tracks, which are far more triangular in outline than normal "Bigfoot" tracks, as well as tape-recording the sounds.

Al Berry was first taken to the site in 1972, and obtained his own tape recordings. Attempts to get photos with trip cameras did not succeed. In 1973 nothing happened, but in 1974 the noises and the tracks were back. I have not heard of anything since. Oddly, with so many sounds on the tape, the high-pitched scream which is usually described by witnesses is missing.

There is a suggested explanation for the whole sequence of events, namely that the first eruption of sound was caused when a creature quietly stealing food scraps upset a kettle of hot water on itself. Having found that during its angry outburst the men did not emerge from their shelter, the creature and its fellows adopted the practice of putting on an intimidating vocal display when the humans were there. The practical problem with that is that if the recording is of sounds that are not typical sasquatch sounds it is of no help to people trying to identify the creatures by the sounds they make.

Sounds are described in almost 10 percent of the sighting reports in my files, but there are a lot of different sounds, or at least a lot of different terms used to describe them. I haven't done a study since 1976, but at that time screams were the only sound consistently reported—37 times —with whistles in second place at 10 reports.

One of the best accounts regarding sounds is contained in a letter I received from Elmer Wollenburg, of Portland, Oregon, describing an experience he had on the afternoon of October 2, 1971, while he was in a park a few miles south of Mount St. Helens, Washington, looking at the scenery on the far side of the Yale Reservoir:

At around 3 o'clock, I heard an unusual sound of tremendous power coming across the lake at me. It started out from a low-pitched raspy quality and gradually raised to a high and clearer pitch. The bellow may have lasted as much as eight or ten seconds. At first I assumed that it was coming from a human being with a bull horn, trying to get an echo from the mountains behind me. It was only gradually that I concluded it could not possibly have been a human voice.

I studied the opposite shore intently for quite a period of time. The air was clear. The sun was more or less at my back. At first I could see nothing moving. The water level in the lake was perhaps 25 feet low, exposing numerous stumps on the beach opposite. I scrutinized these for signs of movement, figuring that one or more might be humans standing or sitting on the beach.

SCREAMS IN THE NIGHT

Within about five minutes, I saw a figure move up from the beach, across the logging road that borders the east side of the Yale Reservoir, and disappear on the other side of the road onto lower ground. A couple of minutes later I saw a second (?) figure following the same route up to the logging road, across it, and then down out of sight beyond.

These figures were of faded light brown, with a grayish tinge, uniform in color all over. They seemed broad in comparison to their height. They moved with an unusual wobbling, almost bouncing, action. It was a movement not at all as humans would walk. I could not distinguish heads. The blobs were all of one color and shape.

At the point where I sighted these creatures on the far shore, there is a canyon leading down to the lakeshore from the highlands above. I saw and heard no logging trucks or other motor vehicles on the east shore logging road all afternoon.

In May, 1972, I went back to Yale Park with binoculars. Looking across the reservoir at the same spot, I saw a yellow Forest Service pickup truck parked at the identical place where the sasquatch (?) had crossed the road and disappeared up the canyon. The truck was almost invisible to me without the binoculars. Is the sasquatch larger than a pickup truck? Lighting conditions were different on the two days. I do not think I should draw any hard conclusions.

Another report with an interesting reference to sound came from Wayne Thureringer, who then lived at Kent, Washington, in 1970. He said that he and another youth saw a sasquatch on the trail between Sunrise Point and Mystic Lake, 7,000 feet up in Mount Ranier National Park, on July 9 of that year. Here is part of what he told me in a taped interview made six days after the incident:

It was about three o'clock in the afternoon and the climb had been somewhat steep, and as we came to the end of one switchback and about to head around to another one we heard a noise. It was like somebody blowing into the neck of an open empty bottle, and we couldn't figure out what it was. It was loud, and we could tell that it was between us and the switchback we had just come up on, so we weren't really fatigued and we decided we would try and look into it more if we could, so we went part way back down the trail towards the noise and we found a log that had fallen crossways across the trail and wedged itself in between two trees and about three feet off the ground that extended down into the bush. So we climbed out on this log without our packs on and

watched at the end of it. We watched for approximately five minutes and we didn't hear anything. We were about to go back when we heard this noise again.

We turned around and we saw this animal emerge from the dense underbrush and walk across, across country from us, probably 30 yards away. We were heading in a southwesterly direction, I believe, and it was headed in a northeasterly direction, completely opposite of us, but parallel at the same time.

Q—What did it look like?

A—It was large...heavy built, hair over its entire body except the insides of the hands and the middle of the fingers and the feet about the middle of the ankle down.

Q—How did it walk?

A—It walked in an upright position, but somewhat slumped over at the waist and bent at the knees.

Q—Any idea how high it was?

A—I would say about seven and a half to eight feet.

Q—Could you see its face?

A—At one point when it turned towards us momentarily to push a bush out of its face, we saw it then...We saw a face that was somewhat rounded, or a very flat face, and actually all I saw of it was the nose and this is the only thing that really caught my attention for that span of time. It was very flat and pushed in.

Q—Would you say it resembled a person, or an ape, or a bear, or what?

A—I would say more an ape.

Q—How long were you watching it?

A—We watched it for approximately 10 to 15 minutes.

Q—What did it do in that time?

A—It didn't do anything, it just kept on walking.

Sasquatches may do a lot of screaming, but screams don't necessarily mean a sasquatch. If they did Puyallup would be the sasquatch capital of the world. There was, and perhaps still is, something unknown that did an awful lot of screaming there. Puyallup is in Washington, just east of Tacoma. The screams were most often heard in an area of mixed woods and subdivisions southeast of the town, but in a lot of other places round about as well. They first came to public attention in July, 1972, when a resident of a new subdivision called Forest Green wrote to the Tacoma *News Tribune* about hearing loud screams one or two nights a month in the woods behind his home He found that his neighbours had not heard the noise and some of them did not believe him, so he finally made a tape recording of it. In the next year he made several such

128

SCREAMS IN THE NIGHT

tapes, but did not bother about keeping them and some were dubbed over. The best, and the only one made in daylight, was copied by other people but the original is gone. It was made, as were most of them, by hanging the microphone of his recorder out a window of his home.

Whatever does the screaming also causes all the dogs in the area to bark frantically, so the sound of the dogs competes with the screams on the recordings. The screams are by far the louder noise, but since the recordings are made from houses there are always dogs a lot closer to the microphone. On this one tape, perhaps because it was daylight, the dogs are silent; but it was made at the time in the early morning when all the birds are making as much noise as they can. Fortunately they are a lot less noisy than the dogs. The unidentified noise is not an "eeeee" scream, but more of a long "whooOooOooOoo" or "woopwoopwoop" at a high pitch and with immense volume. Heard from a distance it can be compared with the sound of a siren far off, but it certainly is not that.

There is no proof that the source of the screams is a sasquatch. That has been inferred from the fact that there have been at least eight reports of sasquatches seen in the same area since the screaming was first noted, plus a good set of 14-inch footprints found crossing a new road, in January, 1975, not far from Forest Green. Two of the sighting reports were by a state trooper on patrol in the area at night.

I spent one evening at Puyallup with two *News Tribune* reporters who told me they had heard screaming at close range the first time they went to investigate the story. I was not nearly so lucky, and have heard it only once in many nights spent listening, although the frantic barking of the dogs at some time during the night is a frequent occurence. Just about all the sasquatch investigators in the Pacific Northwest have spent some time at Puyallup and many of them have heard screams, although none that I know of have succeded in getting a tape recording.

Other than the sasquatch the chief suspect regarding the screams is the coyote. The screams are certainly not normal coyote noises and it is hard to imagine a coyote generating such volume. However there are a number of intermediate noises that some people identify as coyote sounds while others think they are made by the screamer. I don't know what it is that screams, but it is certainly something and I find it absolutely fascinating that it could go on year after year without anyone being able to prove what was screaming. Any reasonable person would assume that it would be impossible for anything to make a noise night after night that terrified people in their beds (although others were never awakened by it) without law enforcement people, or wildlife personnel, or somebody doing something about it. The example at Puyallup proves that to be a mistaken assumption.

More recently a somewhat similar situation has developed in an area north of Snohomish, not far from Everett, Washington. Tape recordings are not good enough to be sure but the sound could easily be made by the same thing, or the same sort of thing, as at Puyallup, and the pattern of something that screams moving through the woods in an area where there are also quite a few homes is just the same. At Puyallup the steady growth of the subdivisions was bound to put a stop to the activity sooner or later, but that does not seem likely to happen in the new area, so perhaps someone, someday will find a way to prove what it is that screams in the night.

Good tracks, 16 inches long, were found in Washington in the fall of 1976, not far from Mount St. Helens. This cast looks very similar to a Bluff Creek "Bigfoot" print.

Tracks in the Snow

I generally tend to downgrade reports of tracks in the snow, because snow is a very changeable material, and is often soft enough so that there is no problem making imprints of any size desired. There are exceptions, however, one of the most notable of which was a set of tracks seen by several elk hunters on Coleman Ridge, near Ellensburg, Washington, on November 6, 1970. Those tracks were studied by more than one hunting party, and reports and pictures of them have come from more than one source. I have a letter from one of the hunters, Oscar Hickerson Jr., of Renton, Washington, giving a brief description of the tracks, but there is a better account in a letter which Nick Carter, an old friend of Mr. Hickerson's as well as a sasquatch researcher, wrote to George Haas. The following is an excerpt:

Oscar is about 55 and has spent most of his recreation time in the Washington, Idaho mountains and wild areas all his life. He had read about the California Bigfoot sightings in True magazine but had never seen anything like that line of tracks in the snow before. At first he, and his hunting partner, Mr. Jess Helton, thought it was some sort of hoax, but a little investigation convinced them otherwise. Both are experienced hunters and trackers.

The two men had gone to a campsite, previously selected, in Hunting Unit 4 E on a Washington Game Dept. map. The spot is about 30 miles N.E. of Ellensburg, not very far east of a game preserve. They followed Coleman Creek up to their camp, which was on Coleman Ridge. They pitched their tent on Nov. 5 and intended to spend the 6th cutting wood in preparation for elk hunting which opened Nov. 7 at dawn. During the night of Nov. 5-6 some six inches of fresh snow fell, the first of the year. It stopped about 5 a.m. Nov. 6.

They woke up, knocked the heavy wet snow off their tent where it was sagging in, made breakfast and were getting out the chain saw when a man, name not given, came up from his camp about 100 yards farther down the ridge. He was all excited about something that had made big tracks through the camp along a somewhat circular route. He was scared, or somewhat upset, for the tracks went right alongside his camper truck and he thought the thing had looked in his windows. From the lack of new snow in the prints they must have been made between the time the snow stopped, at about 5 a.m., and when he got up about 7

a.m. Then he went back to his camp, also to cut wood.

Oscar and Jess sawed down-logs and talked about the news. They thought then it was some joke, but about 11 a.m. the took a break and went over to take a look. The sun was out and it was just above freezing, not thawing very rapidly. They saw the tracks, made by something walking on two legs, and measured some with a yo-yo tape. This is Boeing slang for a six-foot roll-up tape measure. (Both Mr. Hickerson and Mr. Carter worked for Boeing Aircraft Co.) The tracks measured 17 inches long and 9 inches wide. Oscar said he could stand with both his boots on one track touching each other and the two boots together did not reach

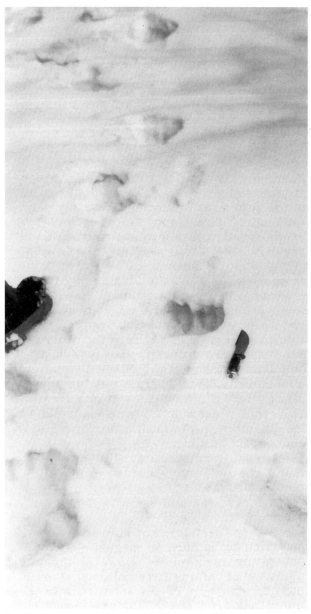

This is one of the pictures that Oscar Hickerson took of the tracks on Coleman Ridge.

130

TRACKS IN THE SNOW

the outside edge of the track across the ball of the foot.

He was impressed by the depth the tracks sank into the snow and by how hard packed the bottom was. Like packed ice, and very flat, no arch showing at all. At first glance he thought about boards on a man's feet but the stride was far too long and they sank too deep. A second glance showed toe action, snow kicked up as the thing stepped ahead, and some heel drags behind individual prints.

In the men's camp area the tracks were fairly close together, 30 to 40 inches, and Oscar got the feeling the Bigfoot had been looking around, moving slowly, for the stride widened beyond the camp as it strode off normally. They followed and came to a place where Bigfoot had stepped over a down log without disturbing the ridge of fresh snow on top. Oscar and Jess looked carefully for some signs of tracks going around the ends of the log, still thinking hoax, but there were no marks at all in the fresh snow except the one line of prints right over the log. He and Jesse had to crawl up over the log where Bigfoot simply stepped over. Oscar is 5'9½" and Jesse 6'1" tall. They figured Bigfoot must have had legs five feet long to make that step. His description of crossing the log was "like getting on a horse" and there were several more along the track farther on.

At that time they followed the tracks for about half a mile, zig-zagging through the woods, curving back up behind their own camp. They cut across to the tent, got their rifles and continued following Bigfoot, also looking for elk sign, most of that afternoon, for a total of about three miles, with all the curves and zig-zagging, meandering around etc. to where the tracks finally went down out of the snow into Coleman Canyon. He said that at times Bigfoot idled in one place, back and forth, then would stride out for a way, then idle around some more. The tracks did not circle but switchbacked around contours. Full stride was measured at between four and five feet, length between tracks varied, some were four, some five and others in between.

The pictures were taken first, in the camp area. Oscar started down to see the tracks without his camera, but went back and got it, fortunately. He took the three shots around 11 a.m. before they tracked Bigfoot into the woods....

Oscar tried to match stride where prints were close together, and it took his maximum stride to "almost" reach. When Bigfoot really stepped out it took two of their strides to match one of his....

The one thing about those tracks which astonished Oscar most, and impressed him deeply, was the flat bottom of those huge feet, and how hard the snow was packed.

In his letter to me, Mr. Hickerson said that if he had been wearing something on his feet the size of the prints it would have been like wearing snowshoes, and there would have been very little depth penetration in comparison with the tracks. Normally with tracks in the snow there is no way to be certain of the condition of the snow at the time the tracks were made, so that it is impossible to judge how much weight was involved in making them, but there would have been no such problem on this occasion. In any event there would have been no way a man could have cleared the fallen logs without breaking stride or disturbing the snow on top of them.

Another impressive set of tracks in snow was seen and photographed in 1934 by Dave Zebo, who was aviation director for Humboldt County, California, at the time when he was interviewed by a correspondent for the *Humboldt Times* at Eureka in 1960. He said that while living at Weaverville he obtained permission from the Forest Service to spend a night at the lookout on top of Mount Bally, south of the city, and set out on an 11-mile ski trip trip to the top of the mountain. The story continues:

Two miles above the timberline, Zebo ran into strange tracks in the snow. There was no animal or human to be seen within range. He stated, "I have never seen anything like these indentations of tracks before or since". The tracks were deep and heavy, but the spacing was what especially drew his interest. The tracks were from 4 to 6 feet apart. Too far for the stride of a normal man, but they were single tracks of a two-footed person or creature.

Pointing to the human element, Zebo said, was the fact that an animal will meander. A human, usually takes a straight path, (and sometimes the hardest way) to his objective; while an animal is known to meander to find the easiest direction.

The footprints in the snow, of which Zebo was so curiously engrossed that he took photos of them, went from the bottom of the mountain to the top, from west to east; there was no deviation at all.

"I followed the old trail, and as far as I could see I saw the tracks, making a single line," Zebo said. "There were no other tracks around and I stayed the night in the lookout and came back down the next morning. A heavy snow fell during the night and covered the tracks."

TRACKS IN THE SNOW

The photos gave Zebo proof that the experience really had happened, and upon returning to Weaverville, he had the pictures developed. He showed these to a number of persons in the vicinity.

Speculation ran high, but no one came up with a solution, nor among those contacted had anyone ever seen such an incidence. The forest personnel were among those contacted, with no better luck at a solution. Everyone was interested and intrigued, and discussed the event for days without solving the mystery. "In those days," Zebo said, "we had not heard of Bigfoot." He has since wondered if Bigfoot was the answer to the puzzle.

To summarize the experience: The big tracks were definitely there. They were single, as a human's would have been, but too wide apart in stride (they never hesitated but went energetically up the mountain, as if made by a creature with gigantic strength) for an average man's....

When I spoke to him on the telephone in 1970, Dave Zebo said the tracks seemed to show considerable weight, sinking four to six inches in hard snow that would have supported the weight of a small animal. No toe prints were discernible. The tracks went on down the other side of the mountain.

Another man who observed strange tracks in the snow long before Bigfoot was ever heard of was Jim Atwell, of Skamania, Washington. He wrote to me in 1969 describing something that had been on his mind for more than 40 years:

In 1927 and 1928 the city of Port Townsend let a contract to Coyne Construction Co. to lay a large water line from Port Townsend, Washington about 28 miles through rough foothills to the upper Quilcene River.

I subcontracted backfilling part of this ditch or covering the pipe in other words. I used a gas power shovel.

Either December of '27 or January of 1928 we had a light snow of approx. 3 inches. This stopped our work but at that time I was 25 years young and payments to meet of $1,000 per month so I hiked the several miles to the water line to check on the equipment. This morning on reaching the water line via a small trail, I crawled up on the pipe for easier walking, the pipe was wooden with metal rods around it every few inches but about 30 inches in diameter and easy to walk on top of.

Soon after I started along the pipe a set of tracks appeared ahead of me in the fresh snow. These tracks came out of a heavy stand of timber and downfall, near where the trail was, never once using the trail. I inspected the tracks on the water line and they appeared to be made by a barefoot man. I did not measure the tracks but would guess from memory that they were made by a large man and about a number 10 or 11. I had never heard of the Abominable Snowman at that time so just guessed that it was some nut of a mountain man that one might find around Quilcene. This bare foot track walked down the pipe about 100 yards and on leaving the pipe he or she jumped about 6 feet across the ditch and landed on a 12-inch log covered with snow, something no logger could have done with caulked shoes, and then headed up the mountain through the roughest kind of going, downfalls, brush

Tracks with a four to six-foot stride photographed by Dave Zebo in 1934 on Mount Bally

TRACKS IN THE SNOW

and rocks all covered with snow.

I went on to where my power shovel was and returned back to town on a different trail. The following day I was down town and I mentioned this track and another man said that he had also seen this same track. I have often thought of this track since then over the 42 years but have never mentioned it much because any listener would wonder what I was drinking or dreaming.

Humans often do crazy things, but this track appeared a little inhuman or more animal.

First it avoided the trail and chose very hard country to travel over. After reaching the pipe it did choose easy walking for 100 yards, then it again chose the roughest going that was leading into the mountains without a house or road for miles.

The snow gave evidence that this was not a hoax. No others around any place within miles and a hoaxer would need to get his shoes back on quick and then he would leave telltale tracks.

Being young and anxious to make my business pay and just thinking one finds bigger nuts every day and most of them are harmless....I missed learning something. I was unarmed and anyway I just wasn't interested in following the tracks up over such rough country....but I should have inspected the tracks where it came from. Nowhere did I see where it came down out of the mountains. It must have come before the snow, which could have been a day or two.....

To get to salt water shore he would have had to go through town and had he dug clams he would have left tracks in the snow. Unless after filling with clams he slept in the brush and woke up after the snow fell. I now kick myself for not backtracking the tracks because they were between the pipeline and town and the distance was not too great even in heavy brush.

These tracks were not near as large as some of the reports indicate....the tracks were not too wide apart as I walked down the pipe between them except when it left the line over onto the log. This was a far greater step or jump than I could make, it left the pipeline across a ditch of about six feet and landed on the small log two feet higher than the pipe and did not mess up the snow but landed neatly. This impressed me more at the time than the barefoot tracks in the snow did. He or she did have powerful muscles.

Probably the most famous and most controversial tracks in the snow were those found near Bossburg, Washington on December 14,

Grover Krantz examining a crippled track that had been preserved by covering it with a newpaper.

1969, by Ivan Marx and Rene Dahinden.

The tracks originated at the edge of Lake Roosevelt (the Columbia River impounded behind Grand Coulee Dam), ascended a steep bank, crossed a railway, a highway and a wire fence, entered the trees, wandered around and eventually returned to the lake.

The tracks were familiar, having been seen in the dirt at a garbage dump not long before. The left foot was normal, 17 inches long and seven wide. The right foot was grossly twisted and had only four toes. Rene spent most of the day examining the tracks and was satisfied that they had been made by living feet capable of toe movement.

These were also the tracks that impressed Dr. Grover Krantz and Dr. John Napier, two physical anthropoligists who concluded after studying casts of the right foot track that it was just too good a job, anatomically, for anyone to have faked.

The problem with the tracks is that some months later Ivan started showing a movie of what he said was the cripple-footed sasquatch. It created a considerable stir at first, but in the end the concensus was that it did not show any such thing, and a cloud was cast over everything else that had gone on in Bossburg.

There had been a report of a similar track seen in that area well before 1969, and there has been one since, but with no photos or casts to prove if it is the same track.

Flurries of Action

The majority of reports of sasquatch sightings and tracks are haphazard in location and apparently unrelated to any other report, but from time to time there are flurries of activity, or at least of reports, from a limited area. Several factors are probably involved in this. When one story brings attention to the subject people who have been keeping quiet about something they saw are more likely to speak up; other people may get started actively digging up information or going out to look for the creatures, a lot of people may just be a bit more on the alert and taking an extra look around, while still others may be trying to get in on the excitement by making things up.

With so many possible reasons for a sudden increase in information it is always difficult to judge whether the sasquatch have anything to do with it at all. Perhaps one or more has come into an area where there were none before, and presumably the chance of an accidental sighting might be better with an animal not yet familiar with its surroundings. Perhaps animals that have always been around have started behaving in a way that attracts attention, or perhaps it is the humans that have moved into a new area, as at Bluff Creek, or who are doing something different.

Looking at it from the point of view of a single researcher maintaining a file of information, it may just be a matter of establishing a contact with people in an area where there has been information all along. Sometimes such contacts are temporary, so the record will show a brief flurry of reports when actually the situation in that area has been about the same before and since. Looking back, and without studying the files at all, I can recall bursts of activity near Klemtu and Bella Coola in British Columbia; Nordegg, Alberta; Bellingham, Colville, Pacific Beach, Puyallup and Stevenson in Washington; Astoria, The Dalles, Estacada, Albany and Myrtle Point in Oregon; Bluff Creek and Oroville in northern California. In some cases there was a great fuss in the local media at the time, in others not much public awareness at all, and of course a lot more happened in some of those places than in others.

Aside from Bluff Creek, which is in a class by itself, Colville and Oroville in 1969-70 attracted the maximum concentrations of sasquatch hunters, but 10 years later they don't look particularly important in the general scheme of things. Stevenson, The Dalles and the area above Estacada are all within about 50 miles of each other, so they could perhaps involve the same population of animals, provided they can swim the Columbia. If that is considered as a single area then it is far the most productive of any as far as sighting reports are concerned, with more than 40 altogether. These come from a great variety of witnesses, and independent reports have come in fairly steadily both before and since the periods of high activity.

The first thing that drew attention to that part of the country was a letter to the editor of *Argosy* after that magazine published stills from the Patterson movie. The letter read:

For several days last June, three friends and I were in constant contact with what is known on the West Coast as a Sasquatch and to the rest of the world as a yeti. There was more than one of them and we took pictures of the footprints, which ranged from nineteen to twenty-three and a half inches in length. The dimensions of the creatures are roughly from eight to ten feet tall, and the weight would be in the neighbourhood of 450 pounds. I want to impress upon you that this is serious business.

I took pictures of this yeti and was as close as five feet to it. The thing stalked us as much as we stalked it.

One more startling fact is that it all took place within two miles of The Dalles, Oregon, a city with a population of 11,000, famed as the end of the Oregon Trail.

D.C.

Roger Patterson made contact with the writer of the letter and the other youths involved. Presumably the pictures did not amount to much, since they have never been mentioned further, but Roger found a situation similar to that at Richland, Washington the summer before, with a group of teenagers spending their nights monster hunting with considerable success. Following is a partial transcript of a tape recording, with Roger talking with one of the boys at the scene of their adventures:

Well, the first time we saw him we were on the road coming from the golf course. There was these corners and it was following us around, you know, it was up on the bluff, following us around the corners. There was rocks rolling down. So we got down, just right over this bank here. He came down... well, we started up the bank first; we thought maybe it was some kids rolling rocks on us or something, 'cause we never did see what it was, we figured maybe it was some kids from the trailer court or something sleeping out. So we started up the bank...

FLURRIES OF ACTION

and as we started up, something started down, you know, and we could see it, and it was pretty big. We knew it wasn't...there are no kids around here that size.

So we went down to that field towards the freeway, we got about, oh, 70 yards into the field, and whatever it was came through the fence. It went through the fence...just went right through it.

So we stopped, and when we stopped so did it. And we looked at it for, oh, about 30 seconds or so. We could see it, you know, it was standing on two feet, and it was big, about eight or nine feet tall, and real heavy. So we decided it was time for us to make it to the freeway and stop some cars and get to town, you know.

So we went on to the freeway, and as we'd run it'd run, and if we'd walk, it walked, and if we stopped it stopped, but when we'd walk his steps were about three times as big as ours and he was catching us. So we decided we'd run. So we ran to the freeway... and as soon as we got there it stopped and wouldn't come any closer to the freeway, so we figured we were pretty safe there, so we watched it again....Then a car came around the corner....the headlights shining down the road towards us, we saw it going down across the freeway. So we decided maybe it was sneaking up on us or maybe not, maybe he just doesn't know anything about us and was just going to the river or something.

They got a ride to town and later returned with some other youths in a car.

It was about four or five in the morning then, it was starting to get light, so we figured we'd come back up here and see what it was in daylight. We got up here, right to where it came down the bank after us, and it was coming back up the bank there. We saw it coming up....So we stopped the car, and everybody grabbed knives, and whatever we could find, we were gonna stone him to death I guess.... We chased him and he came up this bank right here.

They were unable to get close to it, it quickly left them behind.

So we decided that night we would come up with some guns and bring some people up, you know, and gonna track him down and shoot him, you know. So we came back up that night, there was probably eleven, twelve others. We all had pretty big guns, we had .270's and 30.06's, we had a 7 mm., a 6.5 mm....everybody just brought whatever they had and this one kid, D—C— had a 12-guage shotgun. He had 00 buckshot. We went up there, and we saw him the first

night but not close enough to shoot....The second night we came out....there were two of them, and the night before they'd come straight out over toward this other hill. This night they came through the cut and turned right.....so D—C— and I took the 12-guage shotgun and went down to see where they went...

We were walking back up, we walked by this big tree, and D— wanted to check inside the tree, because the branches on the uphill side of the tree came clear to the ground and he wanted to look in there.... I reached down and pulled up a branch, and D— was standing on my right, and we saw it. It was about eight or ten feet away from us, I don't know how far it was, but it wasn't very far. It was in a crouching position. It was downhill from us, and I'd say it was at least nine feet tall when it was standing up, when it was crouched it was probably seven feet tall, or something like that, you know.

So D— let him have it twice with the 12-guage from the hip, just shot him twice That knocked it down, it rolled over twice, and got up and went right through a four-strand fence and snapped three poles off, big wooden poles, snapped them off flush with the ground. So D— and I figured, well, we shot it, it was just gonna roll, and when it stopped rolling it'd be there dead, you know, so as soon as he started rolling we started down the bank after him.

Then when he got up and ran, he ran away from us, so we decided well, he's not as hurt as we thought he was, and we only had one more shot, and we didn't want to go after him with just one shot, so we started back up where everybody else was..

We decided he was wounded and he was more than likely gonna be mad, so the best thing we could do was get off the hill, go back into town and let him cool off and come back up, you know, in the daytime, and scout around, see if we can't find him...we figured on tracking him to wherever he was going.

We figured that, if nothing else, you know, he'd just be running until he couldn't run any more, you know, like sometimes you might shoot a deer or something and he'll go for a hell of a long ways and then you'll find he's dead. So we decided if we came up there in the day time, about noon or whatever, you know, and track him by blood and tracks, we figured there'd be lots of tracks, you know. So we came back up the next morning and we went down and took pictures of where we saw him and the fence he went through and everything, and we tracked him for maybe 80 to 100 yards and

FLURRIES OF ACTION

then we lost track, couldn't find any more track and there wasn't enough blood to follow.

They came back several more times and saw one of the creatures at a distance. They noticed that they could spot them by the red glow of their eyes, and that they could hear them coming, walking among the rocks. Then one night that four or five of them were there they split up and one of the creatures walked up behind a group of three boys.

One of them walked up within eight feet of them....it was in back of them. They didn't even know it was there until it turned around and started to leave. Apparently he was watching to see what they were doing.. ..anyway, they heard him, he kicked a rock or something, and the only gun..(the had) ..was a .22.

R— didn't shoot at him....he figured, you know, he'd just as soon let him leave.

The boys did not go there again after that. They said the creatures left footprints a foot and a half long and eight to ten inches wide. Hair all over the body was dark, reddish brown. Ears, nose and mouth and eyes were like a human's. No snout.

So we ruled out bears, well we ruled them out in the first place because bears don't walk on their back legs and don't run on their back legs and whatever this thing did.

....It smelled rancid, like somebody who hasn't taken a bath for maybe two or three years.... smelled terrible. I don't know what I never smelled anything that smelled like it before so I wouldn't know what it smelled like. It smelled like maybe a cow that's been dead and mouldy for a couple of months, that's about what it smelled like. Terrible.

The boys said that the creatures came down from the hill above the road about 11:30 p.m., and they would station themselves with their guns and wait. Sometimes they saw one, sometimes two, and on one occasion three.

So we'd hear them coming and we'd position ourselves where we could get them in a cross-fire or, well, we'd have him in any kind of trap we want to, but we never did kill him. Well, we never got enough people up there that were really interested....most everybody that was up there was scared to death, you know, including myself.

That is the kind of story one is inclined to reject out of hand, if only because it seems unbelievable that a mob of teenagers (once as many as 13, and one as young as 13) would be allowed to shoot up the coutryside with rifles in the middle of the night. But something very similar had happened at Richland and no one had done anything about it there. In any event the boys' story received some very substantial support when there was another flare-up of sasquatch activity at The Dalles in 1971.

On that occasion Rene Dahinden was one of those who went to see what was going on. He told me that Joe Mederios, the maintenance man at a trailer park west of the city, was the first to see the creature, when he happened to look up from watering flowers early in the evening of May 27. What he saw, walking in the field across the road, was a grey figure that he first took to be a man. The figure started to descend a rock slope to a lower field closer to the road, then turned back and left. By this time Mr. Mederios had decided it was too big to be a man. He also decided not to say anything about it. The next three days he did not see it, but on June 1 the three owners of the trailer court, who had come up from Portland, all saw the creature out the office window. Mr. Mederios then told them he had seen it earlier, and at 10 that night he saw it again, while people in two cars had a spotlight on it.

The following evening, near dark, Dick Brown and his wife, residents of the court, and Mr. Mederios saw the creature in the lower field, standing under an oak tree. Mr. Brown got a rifle with an eight-power telescopic sight and studied it as it again walked slowly away. He estimated its height at nine feet and was particularly impressed with its muscular shoulders, which he estimated to be four feet wide. The hair that covered it was about eight inches long, white with silver tips. Its hands reached below its knees as it walked with its shoulders slightly hunched.

Its knees were always bent as it walked, and it put its feet down flat, not heel first. It's arms swung "as if it was reaching out and pulling something back." There was a crest on its head. Its eyes seemed deep sunk, but when hit by car lights they reflected "like silver dollars." Mr. Brown put a shell in the chamber but decided not to shoot.

By the following day the word was out and the area was overrun with people. Many hundreds of man hours were spent watching in the area after that, but the show was over.

Most of the activity on the other side of the Columbia was between the dates of the two spells of excitement at The Dalles.

FLURRIES OF ACTION

In the month of March, 1969, a sighting was reported on the road near Skamania very early in the morning. It was an ordinary sort of report but for some unknown reason it caused a wave of publicity and set a lot of people searching. The result was the discovery of two sets of giant tracks well back in the hills, one of which was duly cast by the local sheriff. Further sightings were reported in March and November, and track finds in April and October. In 1970 there were more tracks found in June and another sighting in August. The last was probably the most interesting. Mrs. Louise Baxter, of Skamania, was out of her car checking for a flat tire when she saw a creature standing in the wooded area by the road with its left fist up to its mouth as if it were eating something. She described it as coconut brown, shaggy and dirty looking, with a big head set right on its shoulders, a jutting chin and a receding forehead. The nose was wide with big nostrils. The eyes were amber and seemed to glow. She couldn't recall getting back in the car and getting it going, but "as I pulled out I could see it still standing there, all 10 or 12 feet of him.

Probably the most interesting situation of any is that at the mouth of the Nooksack River near Bellingham, Washington. In 1967 when the sockeye salmon were running in unusual numbers in the river there were so many sasquatch sightings that no-one has a complete record of them. Nobody had to go looking for the sasquatch, apparently, the problem was that they wouldn't get out of the way. Although I lived only a couple of hours' drive away I didn't hear about that situation when it was going on, but I did spend time looking into it a couple of years later. The big sockeye runs come only once in four years, so in 1971 everyone was waiting to see if there would be a repetition of the sasquatch activity, but there wasn't. Then in 1975, with everyone off guard, it happened again. That time Rene Dahinden moved into the area for a while, and Jon Beckjord, just beginning his participation in the sasquatch search, came up from San Francisco and stayed. Most of the land involved is Indian property, and in 1975 the Indian "law and order" organization kept control of the situation, conducting police investigations on many of the incidents.

The Nooksack River gets its start in life on the slopes of the highest mountains in northwest Washington, but it runs about 20 miles through flat farmlands before it gets to the sea. There is an area of heavy forest on the Lummi peninsula, although it is cut up with roads and there are many houses. There is also heavy growth, and no roads or houses, on the islands in the mouth of the river. It isn't an area that could be expected to house a population of sasquatches on a permanent basis, but if they used the river for a highway, as the Indians say they do, they could easily come down at night and settle in for the fishing season.

Most of the 1967 sightings took place in September, and more than half of them were by fishermen drifting with gillnets down the channels at the mouth of the Nooksack. Mr. and Mrs. Joe Brudevold told me that they had seen an eight-foot, black animal with a flat face standing in the river in the early afternoon. It was about 200 yards away, and although the water was only up to its knees it bent down and disappeared in it. The river is muddy, so that neither salmon nor sasquatch could be seen beneath the surface, but I was told that sometimes a surge would travel along the river as if something very big was swimming by. In the area of the Brudevold sighting tracks were later found coming out of the river onto a sandbar and covering about 150 yards before re-entering the water. They were 13½ inches long and sank in two inches, compared to ½ inch for human tracks. They were flat, had five toes, and took a 45-inch stride.

Sasquatch and human tracks on the sand of the Nooksack estuary.

FLURRIES OF ACTION

An oddity was that the right foot was twisted outward at a 45 degree angle, but only on every second step. The toes kicked up some sand. These tracks were photographed, but only after the tide had been in and out and a lot of silt had settled in them.

In the same period Johnny Green, drifting down a channel at night, noticed his net pulling out of his boat. He thought it had snagged, but checking the other end with a powerful spotlight he saw a sasquatch standing in the water pulling the net towards itself as if it wished to take a fish out of it. He shouted, and Reynold James and Randy Kinley, who were waiting their turns at the head of the drift, came down. With all their lights on it the creature let go of the net, waded ashore and walked into the trees. I talked to all of them when I was working on a computer survey, and John Green (no relation) later was shown in a movie taking and passing a lie detector test concerning the incident.

Another interesting story was that of Martha Washington, who went out of her house because her dog was barking and saw by flashlight a dark brown creature sitting on an upended oil drum in her neighbour's back yard with its hands on the edges of the drum and its stiff arms pushing its shoulders up. Its eyes reflected the light, but its face was quite human looking. She went back in the house.

All of those incidents were on the banks of the river, or else right in it. There was another sighting, however, by a woman whose house overlooked the sea to the south of the river mouth. She told me that one morning as she was returning from the store she saw a sasquatch wading out of the water and up onto the shore.

In 1975 the sighting reports were not concentrated at the mouth of the river, but tended to be in the bush on the peninsula. Police reports show that in the early part of October there were several calls concerning some creature making loud noises in the vicinity of houses, and there were four reports of hairy bipeds being seen. On one occasion the captain of the police force had fired his revolver at an animal he saw in the bushes beside a field. The thing appeared to be about six feet tall, and covered with hair. Descriptions given indicated that more than one creature had been seen, although not more than one at a time.

On October 23, at 7:30 p.m., the police sergeant answered a call to a house where something had been heard pounding on the back wall. The woman who lived there had gone next door to her son's house and there was no prowler to be found, but something had apparently torn some plastic that covered a back doorway, and there was a window broken. At 2:20 a.m. the same night something was again reported behind the house, and when the sergeant arrived, along with several other people, his spotlight quickly picked up what looked like a very large ape standing in the back yard. While someone else held the light on it the sergeant walked to within 35 feet of the animal, which made no attempt to run but crouched down as he got near. There they stayed for "many minutes," while the sergeant wondered what to do next. He had a shotgun loaded with buckshot but he was not sure if the thing was some kind of human, and if it wasn't, he didn't know how much buckshot it could take. He noted afterwards in his report that it was black in color, would stand seven to eight feet tall and appeared to have no neck. It was covered with short hair, except on the face. He could see no ears. The eyes were small. It appeared to have four teeth larger than the others, two upper and two lower. Its nose was flat. He could see the nostrils. At the end there were seven people watching it, although only two others approached close to it. Then there were noises heard off in the dark at both sides, and the man with the spotlight swung it off to the right and called that there was "another one over there." At that point the sergeant decided to return to his patrol car.

138

Apes and Men

The question that bothered the police sergeant, "is it some kind of human," bothers a lot of other people too, but there is no reason why it should. The answer is no.

Because we don't know any other creature that gets around the way we do, humans tend to identify humanity with their shape. Even when on all fours or sitting or lying down the human proportions that result from our bipedal means of locomotion are readily recognizable. But there is plainly no reason why there could not be some other species of animal, or more than one, that stood and walked like a human without having developed the other attributes of humanity.

It is not really our shape that makes the difference between us and other animals, it is the things that we do. Members of the most primitive of human societies have the ability to communicate complicated information by sound alone, and remember it in the form of sounds, so that one individual can call on the experience of others to let him know what to do in situations that are new to him and in places he has not ever been before. No animal is known to do anything of the sort.

Field studies of wild animals in recent years have produced a wealth of new information and have disproved a lot of what everyone thought was known of animal abilities, but there has been no suggestion that any animal has a language. When it comes to making things, animals such as the spider and the beaver have remarkable talents, but they are inherited, not learned. Vultures and mongooses will throw rocks to break ostrich eggs, and chimpanzees will crush leaves to make a sponge and trim twigs to poke in a termites' nest, but that is about as far as animals go.

In more than 20 years of investigating reports of giant hairy bipeds I have constantly looked for indications as to whether they were human or animal or something in between. The initial impression, from Indian traditions, was that they were some sort of human. Albert Ostman was of the same opinion and some of the things he claimed to observe tend to support that view, but in his descriptions of the actual animals there was no hint of humanity, and the same is true of the assembled mass of testimony. A sasquatch walks in much the same way as a man and therefor has a considerable physical resemblance. It is also like a man and unlike other apes in its omniverous diet, its ability to swim, and its survival in a

wide variety of climates and conditions. That about completes the list of similarities. The list of differences is much longer.

A sasquatch has a fur coat. It has the size and strength to discourage most or all predators. It may have the speed to run down its own prey, and it certainly has the strength to kill most other animals. It can see in the dark. It requires no shelter and no clothing. It does not need tools, weapons or fire. In all these things it is the opposite of man. Most significant of all, it is a solitary animal. It appears most likely that the largest groupings, normally, would be a female and young, or perhaps a family with an adult male. Mankind, for probably millions of years, has depended on numbers for survival. His societies always include groups of families, and the interaction with so many other individuals probably had as much to do with making him what he is as did his upright posture or his use of tools. Speech, co-operative effort and tool use go with the big brain, which is what really distinguishes man from all other animals. All of those things the sasquatch never needed.

Humans have relied on tools and weapons and our skills in using them for so long that we have atrophied physically and could not possibly survive without them. There may be a question whether the development of the human brain was in compensation for physical deficiencies or whether human physical abilities declined because brain power made them unnecessary, but either way our bodies and brains are tailored to each other. No species without such a brain could survive in such a body, and no animal that was physically the master of its environment would have had to evolve such a brain.

In one basic way the efforts of the sasquatch hunters must be counted as a total failure so far. They have not been able to prove that such creatures exist. They have not even been able to convince science that the matter is worth looking into. A conference on the subject was held at the University of British Columbia in 1968, and a good many academic papers were presented, but most of them dealt with the subject in the realm of folklore. Aside from a book containing some of the papers the conference did not have any apparent result. No acedemic institution, so far as I know, has anyone involved in either field research of the compiling of available information.

My own files have grown to such unwieldly proportions that it is not practical to do statistical studies just for the sake of finding out what the current figures are, and years of experimenting, including four years of feeding material into a computer, have demonstrated that the accumulated information is of no use in trying to catch up with one of the creatures.

For those who are keeping score I can say that in May of 1980 I had 365 sighting and track reports from California; 297 from Washington, 245 from B.C. and 182 from Oregon.

Additional numbers have not changed the distribution of reports throughout the year, July and August are still the top months, followed by September and October, and May, for some unexplained reason, remains at the bottom.

Breakdown in May, 1980, was:

J	F	M	A	M	J	J	A	S	O	N	D
45	39	34	49	31	61	90	87	72	74	56	52

Those figures are for the three west coast states plus British Columbia, but adding the rest of North America would only make the figures larger with no significant change in the proportions.

There are still more reports in the daytime than at night, but the difference has tended to narrow over the years. At present, on the Pacific coast, I have 215 reports recorded in the daytime, 167 at night. Since there are far more human observers around in the daylight and they can see much better then, it appears very probable that the sasquatch ' are actually around more often at night.

As to where the sasquatch are seen and who sees them, the most common form of sighting is by someone driving on a side road, in the fall. The second most common is by someone at home seeing something outside.

Both those instances involve massive numbers of potential witnesses, however. On an individual basis there is no doubt that someone out in the bush, fishing, hunting, prospecting or whatever, is more likely to have an encounter than someone driving around or staying at home.

As to what people see, in most cases it is a solitary creature that is presumed to be a male, although it may be that only lactating females have prominent breasts and that otherwise there is no easy way to tell. Reports of more than one individual make up less than 10 percent of the total, and there are only a couple of dozen reports of more than two together. There is no specific count of more than four, although in a couple of cases, such as the group Burns Yeomans watched there may have been more.

Very little has been learned that would be of any use to a hunter. No food has been identified that will attract sasquatches, nor do we know where they sleep or anything about their daily routine. We don't know how they spend the winter. It is obvious from the scarcity of tracks in the snow that they are not moving around much, but whether they hibernate or what else might account for their absence no one knows.

But while there may be nothing that is of use to the hunter there is certainly enough information to establish two very basic points—and in both cases what has been established by observation is the exact opposite of what almost everyone takes for granted.

Point one is that the sasquatch are not endangered. They are reported over such a wide area (far wider than is dealt with in this particular book) that even if there are very few of them in any one place they must still number many thousands; and aside from the destruction of their habitat, which is happening in only a very minor proportion of their range, man is not doing anything to endanger them. Certainly they do not suffer from hunting pressure. There is no proof that man has ever killed even one.

Point two is that the sasquatch are not dangerous. Whatever its historic cause, the instinctive fear that humans seem to have of them is not justified by anything that is known today. The wild areas of most of North America have long been considered safe for humans to wander around in, even unarmed, and if the sasquatch are there now they have always been there. They are big and they are interesting, but they are nothing to be afraid of.

Index

BIGFOOT/SASQUATCH
history, research, evidence, encounters

Bigfoot Encounters in Ohio
C. Murphy, J. Cook, G. Clappison
0-88839-607-4
5½ x 8½, sc, 152 pp

Bigfoot Encounters in New York & New England
Robert Bartholomew Paul Bartholomew
978-0-88839-652-5
5½ x 8½, sc, 176 pp

Bigfoot Film Controversy
Roger Patterson, Christopher Murphy
0-88839-581-7
5½ x 8½, sc, 240 pp

Bigfoot Film Journal
Christopher Murphy
0-88839-658-7
8½ x 11, sc, 106 pp

Bigfoot Research: The Russian Vision
Dmitri Bayanov
978-0-88839-706-5
5½ x 8½, sc, 432 pp

Bigfoot Sasquatch Evidence
Dr. Grover S. Krantz
0-88839-447-0
5½ x 8½, sc, 348 pp

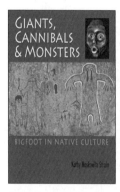

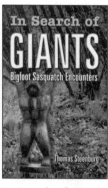

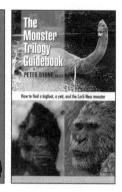

Giants, Cannibals & Monsters
Kathy Moskowitz Strain
0-88839-650-3
8½ x 11, sc, 288 pp

Hoopa Project
David Paulides
0-88839-653-2
5½ x 8½, sc, 336 pp

In Search of Giants
Thomas Steenburg
0-88839-446-2
5½ x 8½, sc, 256 pp

The Locals
Thom Powell
0-88839-552-3
5½ x 8½, sc, 272 pp

Know the Sasquatch
Christopher Murphy
978-0-88839-657-0
8½ x 11, sc, 320 pp

Monster Trilogy Guidebook
Peter Byrne, F.R.G.S.
978-0-88839-723-2
5½ x 8½, sc, 176 pp

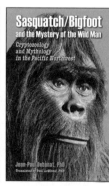

Raincoast Sasquatch
J. Robert Alley
978-0-88839-508-5
5½ x 8½, sc, 360 pp

Sasquatch: The Apes Among Us
John Green
0-88839-123-4
5½ x 8½, sc, 492 pp

Sasquatch/Bigfoot and the Mystery of the Wild Man
Jean-Paul Debenat
978-0-88839-685-3
5½ x 8½, sc, 428 pp

Sasquatch in British Columbia
Christopher Murphy
978-0-88839-721-8
5½ x 8½, sc, 528 pp

Tribal Bigfoot
David Paulides
978-0-88839-687-7
5½ x 8½, sc, 336 pp

Who's Watching You?
Linda Coil Suchy
978-0-88839-664.8
5½ x 8½, sc, 408 pp

Hancock House Publishers | www.hancockhouse.com | sales@hancockhouse.com